Biomass to Biofuel Supply Chain Design and Planning under Uncertainty

Biomass to Biofuel Supply Chain Design and Planning under Uncertainty

Concepts and Quantitative Methods

MIR SAMAN PISHVAEE

Associate Professor, School of Industrial Engineering, Iran University of Science and Technology, Tehran, Iran

SHAYAN MOHSENI

PhD Student, School of Industrial Engineering, Iran University of Science and Technology, Tehran, Iran

SAMIRA BAIRAMZADEH

Postdoctoral Researcher, School of Industrial Engineering, Iran University of Science and Technology, Tehran, Iran

ELSEVIER

ACADEMIC PRESS

An imprint of Elsevier

Academic Press is an imprint of Elsevier
125 London Wall, London EC2Y 5AS, United Kingdom
525 B Street, Suite 1650, San Diego, CA 92101, United States
50 Hampshire Street, 5th Floor, Cambridge, MA 02139, United States
The Boulevard, Langford Lane, Kidlington, Oxford OX5 1GB, United Kingdom

Notices
Knowledge and best practice in this field are constantly changing. As new research and experience
broaden our understanding, changes in research methods, professional practices, or medical treatment may become necessary.

Practitioners and researchers must always rely on their own experience and knowledge in evaluating
and using any information, methods, compounds, or experiments described herein. In using such
information or methods they should be mindful of their own safety and the safety of others, including parties for whom they have a professional responsibility.

To the fullest extent of the law, neither the Publisher nor the authors, contributors, or editors, assume any liability for any injury and/or damage to persons or property as a matter of products liability, negligence or otherwise, or from any use or operation of any methods, products, instructions, or
ideas contained in the material herein.

Library of Congress Cataloging-in-Publication Data
A catalog record for this book is available from the Library of Congress

British Library Cataloguing-in-Publication Data
A catalogue record for this book is available from the British Library

ISBN: 978-0-12-820640-9

For information on all Academic Press publications
visit our website at https://www.elsevier.com/books-and-journals

Publisher: Brian Romer
Acquisitions Editor: Peter Adamson
Editorial Project Manager: Hilary Carr
Production Project Manager: Prasanna Kalyanaraman
Designer: Victoria Pearson

Typeset by Thomson Digital

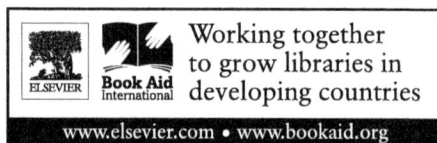

Working together
to grow libraries in
developing countries

www.elsevier.com • www.bookaid.org

Contents

Preface

Growing concerns over climate change, heavy dependence on fossil fuels, increasing demand for energy, and rising oil prices are the main drivers behind the development of renewable energy that would be more cost-effective, less polluting, more efficient, and more sustainable. Among different sources of renewable energy, bioenergy is one of the most prospective sources that could make a substantial contribution to meet global energy demand. Biofuels, as an important source of bioenergy, has attracted considerable attention in recent years for replacing fossil fuels in the transport sector due to numerous merits such as the possibility of production almost anywhere in the world as well as environmentally friendly potential and lower negative impacts on the ecosystem. Investment in the bioenergy industry has been encouraged in many countries around the world by setting national biofuel goals and mandates, with the purpose of replacing specific proportions of non-renewable sources with renewable ones according to a predefined time schedule. For example, the European Council established a binding EU-wide target to source at least 32% of their final energy consumption from renewables by 2030, including a possible upward revision in 2023. However, there are a variety of barriers and uncertainties preventing the large-scale and cost-competitive production of biofuels, and therefore commercialization of the biofuel industry. The main purpose of this book is to produce a complete framework to address the commercialization aspect of biomass to biofuel projects by offering supply chain design and planning models that provide a structure to achieve successful commercialization.

Biomass-to-biofuel supply chains are exposed to a wide range of uncertainties and risks arising from issues such as technology evolution, changing policies and regulations, demand and price variability, unpredictable weather conditions, production cost variations, as well as man-made and natural disasters. Failure to hedge against all such uncertainties may result in suboptimal or even infeasible supply chain decisions. To ensure the large-scale and sustainable production, it is of extreme importance to develop an efficient supply chain that would be reliable enough to function well under dynamic and uncertain business environments for many years. In order to address this problem, this book aims to propose a general framework for biomass-to-biofuel supply chain design and optimization under uncertainty, which can be successfully used for the commercial-scale implementation of biofuel projects by taking into account the problems and challenges encountered in real supply chains. Thematically, the book focuses on the design and optimization of biomass-to-biofuel supply chains with particular emphasis on quantitative methods developed to solve biofuel supply chain problems under uncertainty.

Therefore, the readers of this book may be classified into at least two groups: (1) researchers and students, who can utilize an extensive overview of emerging research challenges and opportunities which is provided by the book in design and analysis of biomass-to-biofuel supply chains, and (2) practitioners and policymakers, who need a flexible platform for commercial-scale implementation; this group can utilize the general framework for biomass-to-biofuel supply chain optimization proposed in this book that incorporates promising biomass sources, different biofuel options, and major production pathways, which can be readily employed for national-level case studies in a large geographical area. The book can also be used as an excellent textbook for coursework or as a self-study and reference guide on two main topics, involving the design and planning of biomass supply chains, as well as optimization approaches to deal with uncertainty in input data of mathematical models. This book comprises two main parts. The first part (Chapters 1–5) sets out to describe key issues related to biofuel supply chains that is organized as follows:

Chapter 1 provides a comprehensive review of biomass feedstocks currently used in the biofuel industry or being investigated as potential sources under three headlines: first, second, and third-generation biomass. The advantages and disadvantages of three generations of biofuels, along with their challenges and opportunities for commercial-scale biofuel production, are also discussed.

Chapter 2 presents a general structure of the biomass supply chain and describes specific activities and operations of the chain corresponding to biomass production, harvesting, collection, storage, preprocessing, conversion, transportation, and distributions. Moreover, a comprehensive overview of biomass conversion pathways from a variety of biomass sources into biofuels and bioenergy products is provided, along with a description of different technologies of biomass conversion.

Chapter 3 proposes a decision-making framework for biomass-to-biofuel supply chains, which categorizes strategic, tactical, and operational level decisions that are made in different stages of the supply chain, including biomass supply and pre-processing, biofuel production, biofuel blending and distribution, and biofuel sales. At the end, this chapter reviews the literature of Biofuel supply chain design and planning problems according to the proposed planning framework.

Chapter 4 proposes a novel risk management framework for biomass-to-biofuel supply chains. In order to ensure that the supply chain model is effective and practical for real-world applications, it must provide an adequate safeguard against uncertainty, and therefore, there is a need to develop a comprehensive risk management framework for supply chain optimization models under uncertainty. To achieve this, the developed framework deals with risks and uncertainties threatening the supply chain in three stages: risk identification, risk assessment, and risk treatment.

Chapter 5 focuses on three dimensions of the sustainability paradigm, namely, economic, environmental, and social sustainability in biomass supply chains. First, it reviews different paradigms that have been proposed in the supply chain management literature, and discusses the economic aspect of sustainability in biofuel supply chains. Then, the four-phase life-cycle assessment (LCA) methodology for sustainability analysis of biofuel supply chains, and characteristics of various life cycle impact assessment methods are described in detail. The chapter ends with the description of ISO 26000:2010, as well as a brief review of social impact assessment credible methods and guidelines.

The second part of the book (Chapters 6–9) focuses on the modeling and optimization of the biomass-to-biofuel supply chain under uncertainty using different quantitative methods in order to determine the optimal design and decisions of the supply chain considering the concepts, problems, and issues described in the first part. The second part is organized as follows:

Chapter 6 provides a comprehensive overview of leading optimization approaches for hedging against various types of uncertainty in the biomass supply chain design and planning models, along with a detailed description of mathematical formulations of the uncertainty modeling approaches. Finally, it classifies and reviews the literature of biofuel supply chain studies according to the source of uncertainty, uncertainty modeling approach, the biomass type, and case study region.

Chapter 7 discusses strategic uncertainties that must be considered at the design phase of biofuel supply chains. In view of the fact that the supply chain design and strategic-level decisions are difficult to change in the short term, this chapter introduces optimization approaches to immunize the supply chain design against these risks. To this aim, a two-stage model is described to show how biofuel supply chains are designed under uncertainty. At the end of the chapter, a case study of the switchgrass-to-bioethanol supply chain is illustrated to exhibit the applicability of the model.

Chapter 8 concentrates on tactical decisions that are made at different stages of biomass-to-biofuel supply chains, and discusses tactical/operational uncertainties in the supply chains. First, to illustrate how to provide an optimal planning model for biofuel supply chains, a multi-period mixed-integer linear programming model (MILP) is presented to address the master planning of Jatropha curcas L. (JCL)-to-biodiesel supply chain under uncertainty. Then, to address the biorefinery process synthesis and design problem, a biorefinery superstructure model for biodiesel production from microalgae is proposed, taking into account uncertainty in technical factors.

This book ends with Chapter 9, which presents the operational decisions that are made at different stages of biomass-to-biofuel supply chains over a short-term, and discusses the most recognized types of uncertainties affecting the operational decisions. Second, in order to address the harvest-scheduling problem, a short-term corn stover harvest-planning model is presented in this chapter, which determines the number of required bailers and assigns the optimal sequence of fields to each baler within its allowable time window. Finally, to cope with the uncertainty in the selling price of stover, a data-driven robust optimization is adopted.

We would appreciate and welcome constructive criticism and feedback from the readers together with suggestions for further improvement of the book for the next edition.

Jun 2020

Mir Saman Pishvaee
Shayan Mohseni
Samira Bairamzadeh

CHAPTER 1

An overview of biomass feedstocks for biofuel production

1.1 Introduction

Growing concerns over climate change, heavy dependence on fossil fuels, increasing demand for energy, and rising oil prices are the main drivers behind the development of renewable energy that would be more cost-effective, less polluting, more efficient, and more sustainable. In order to accelerate the transition toward renewable energy, many countries around the world have set renewable energy targets and introduced regulations and legislation with the purpose of replacing specific proportions of nonrenewable sources with renewable ones according to a predefined time schedule (Yue, Fengqi, & Snyder, 2014). For example, the European Council established a binding EU-wide target to source at least 32% of their final energy consumption from renewables by 2030, including a possible upward revision in 2023 (EC, 2018).

Among different sources of renewable energy, bioenergy is one of the most prospective sources that could make a substantial contribution to meeting global energy demand. Bioenergy is defined as all types of renewable energy derived from biological raw materials known as biomass and includes two forms: traditional and modern. The first form, also called solid biomass, refers to fuelwood, agricultural residues, charcoal and animal waste used in rural areas for cooking and heating through traditional methods such as open fires and cookstoves, and the second form refers to different energy carriers (such as solid, liquid and gaseous energy products, electricity, and heat) derived from biomass at higher efficiencies through technologies and conversion processes such as anaerobic digestion, gasification, and pyrolysis (IRENA, 2014). Bioenergy accounts for almost two-thirds of the world's current renewable energy demand, with more than half of that coming from modern biomass. It is envisaged that solid biomass would be phased out gradually and the demand for modern biomass would grow dramatically.

Biofuel, an important source of bioenergy, is considered as an attractive alternative for replacing fossil fuels in the transport sector due to numerous merits such as (1) possibility of production almost anywhere in the world from biomass helping the security of energy supply, (2) environmentally-friendly potential and lower negative impacts on the ecosystem, (3) renewability biodegradability, and contribution to sustainability, (4) developing agriculture and its related industries, and (5) helping the growth of rural areas and employment opportunities (Simionescu, Albu, Raileanu Szeles, & Bilan, 2017; Kamani et al., 2019). Two forms of biofuels in the transport sector are liquid and gaseous fuels. The advantages of liquid biofuels such as bioethanol and biodiesel over gaseous biofuels (i.e., biogas) such as biomethane and biohydrogen are their compatibility with existing vehicle engines without requiring any major modifications, as well as their ease of transportation and storage (Guldhe et al., 2017). At present, liquid biofuels account for a small share (about 3%) of the global transport energy use. The market size for liquid biofuels is anticipated to grow significantly in the upcoming years. The target set by the International Energy Agency (IEA) is to meet more than a quarter of the world transportation fuel demand with liquid biofuels by 2050 (Pandey, Bajpai, & Singh, 2016).

Identifying the most suitable biomass resources is considered as one of the most important issues regarding sustainable biofuel production in the future. In general, the salient characteristics of an ideal biomass feedstock are: (1) low production cost, (2) high yield (i.e., maximum biomass dry matter per hectare), (3) high energy content (i.e., maximum biofuel production per biomass unit), (4) high output-to-input energy ratio (i.e., efficient use of light, nutrients, and water), (5) rapid maturation and long life span, (6) low water, energy, fertilizer requirements, (7) ability to grow on degraded or marginal lands, (8) adaptability to different soils and climates, (9) resistance to stress factors such as low or high temperature, nutrient deficiencies, salinity, flooding and drought, (10) minimum environmental containment such as pesticides, chemical fertilizers, and heavy metals, and (11) all year round availability and flexibility in harvest time (Anderson et al., 2011; Langholtz et al., 2014; Nanda, Ajay, & Kozinski, 2016). Depending on the type of biomass feedstock used for their production, biofuels are categorized into three classes: first-, second-, and third-generation biofuels. First-generation biofuels are produced from agricultural crops that are in competition with the food industry such as starch, sugar, and vegetable oils. Second-generation biofuels are made from nonedible materials such as agricultural and forestry residue, dedicated energy crops (e.g., jatropha) and industrial and municipal waste. Third-generation biofuels refer to those derived from microalgae (Saladini, Nicoletta, Federico, Nadia, & Bastianoni, 2016).

1.2 First-generation biofuels

First-generation biofuels are obtained from biomass feedstocks that are also used as main sources for food and feed production. Various biomass resources have been utilized for the commercial production of first-generation biofuels around the world. Based on their main content as a determining factor in the type of biofuel produced, these resources can be categorized into two major groups: sugar/starch feedstocks and edible oil feedstocks, which are generally converted to bioethanol and biodiesel, respectively (Ho, Huu, & Wenshan, 2014).

1.2.1 Sugar/starch feedstocks

Sugar crops (e.g., sugar cane and sugar beet) and starchy crops (e.g., corn, wheat, and sorghum) are two main sources for first-generation ethanol production. The overall process of bioethanol production consists of three stages: (1) preparing the solution that contains fermentable sugars, (2) fermenting sugars into ethanol using yeast species such as *Saccharomyces cerevisiae*, and (3) separating and purifying the produced ethanol usually by distillation–rectification–dehydration (Vohra, Jagdish, Rahul, Satish, & Patil, 2014). Although these stages are almost common in the conversion of the two sources into ethanol, the obtainment of fermentable sugars from starchy crops is more complex and energy-demanding than that from sugar crops (Zabed, Sahu, Suely, Boyce, & Faruq, 2017). This complexity lies in the fact that sugar crops can easily be converted to fermentable sugars by the extraction process, while starchy crops must undergo hydrolysis (mostly with the help of α-amylase and glucoamylase) to beak glycosidic bonds connecting the glucose units of starch, producing simple sugars for fermentation (Bušić et al., 2018).

Today, corn is the largest source for bioethanol production, accounting for about 60% of the global ethanol supply. The availability of corn in many parts of the world, maturity of corn to ethanol conversion technology and low production cost of corn-based ethanol have made corn as a major biomass source for commercial ethanol production. The dominant supplier of ethanol in the world with nearly 60% of the total global production is the United States, where about 95% of ethanol production comes from corn starch (Mohanty et al., 2019). There are two conventional methods for processing corn, namely, dry-milling and wet-milling. In the dry-milling process, the whole corn kernel is ground into fine particles by hammer mills before passing through hydrolysis, fermentation, and distillation processes.

In the wet-milling process, the corn kernel is soaked in an aqueous medium and fractionated into its basic components such as starch, gluten, fiber, and germ which are processed separately to produce ethanol and valuable co-products such as high-fructose corn syrup (HFCS), glucose syrup, and dextrose (Vohra et al., 2014).

Sugar cane ethanol, which is recognized as an economically viable biofuel, constitutes around 25% of global ethanol production. Sugar cane has been one of the most competitive ethanol feedstocks in the last decade due to the availability of well-established technologies and infrastructures for its cultivation and conversion as well as the various desirable features of sugar cane such as high fecundity, fast growth rate, and high sucrose content that can be easily converted to ethanol (compared to starchy crops that require preprocessing before fermentation) (Srirangan, Lamees, Murray, & Perry, 2012). However, the cultivation of sugar cane is restricted to tropical and subtropical regions because it thrives in warm and humid climates and has limited cold tolerance (Friesen, Murilo, Florian, Daniel, & Rowan, 2014). The main feedstock for ethanol production in Brazil, the second-largest supplier of bioethanol, is sugar cane juice that is responsible for almost 79% of its production (Sanchez et al., 2008).

With the share of almost 8%, molasses is the third major source being used for producing ethanol in the world. It is a by-product obtained during processing sugar cane and sugar beet in sugar refineries. Approximately 75% of the global molasses is supplied by sugar cane grown in tropical regions of Asia and South America, while the remainder is supplied by sugar beet grown in temperate regions of Europe and North America (Mukhtar, Asgher, Afghan, Hussain, & Zia-ul-Hussnain, 2010). Despite containing a considerable amount of readily utilizable carbohydrates in the form of fermentable sugars, molasses carries the risk of shortage since its production rate is heavily governed by the conditions of sugar cane and sugar beet production (Van der Merwe, Cheng, Görgens, & Knoetze, 2013).

There are many other biomass feedstocks that can be used for first-generation ethanol production, including wheat, cassava, yam, potato, cassava, oat, barley, rice, etc. These feedstocks have a relatively small share of global production and are expected to remain so for the foreseeable future due to several reasons. One of the main reasons is that some of these crops such as wheat and rice have been for centuries the staple foods for millions of people in many parts of the world, making them socially unacceptable for biofuels purposes (Srirangan et al., 2012). Another important reason is

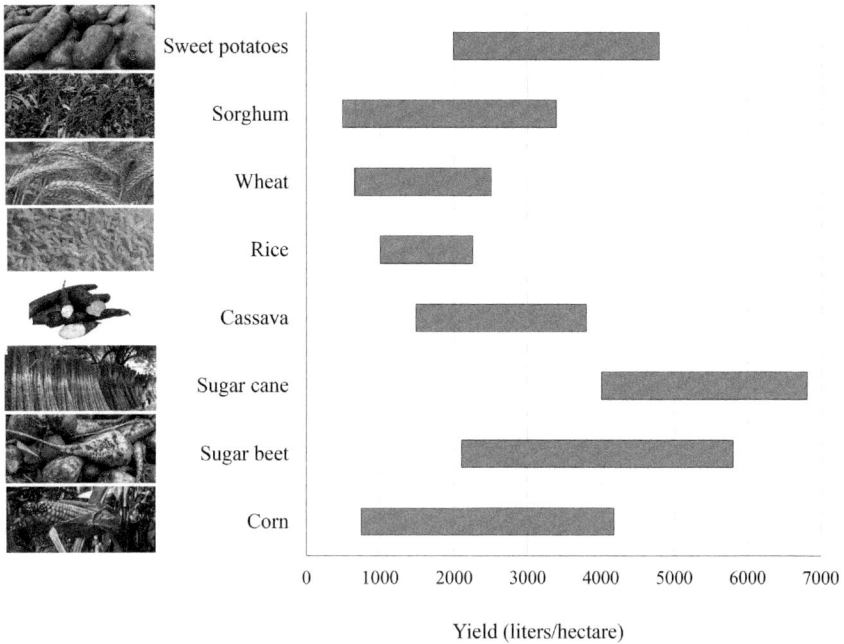

Figure 1.1 *Range of bioethanol yields from various crops.* Based on data from Goldemberg et al. (2010), Leal, Luiz, Horta, and Cortez (2013), Johnston, Jonathan, Tracey, Chris, and Monfreda (2009), Rajagopal, Steven, David, and Zilberman (2007), Naylor et al. (2007), Zabed et al. (2017).

that the productivity (i.e., ethanol yield per hectare) of most of these crops is lower than that of corn and sugar cane (see Fig. 1.1). There are also problems specifically associated with each crop. For example, the cultivation and harvesting of cassava are labor-intensive that lead to high production costs, and the root of cassava is highly perishable after harvest, making it economically and technically uncompetitive with corn (Ziska et al., 2009).

1.2.2 Edible oil feedstocks

Edible oil feedstocks are another important source of biomass with a high potential to produce biodiesel as a promising substitute for current petrol-derived diesel. Biodiesel is currently the second most produced biofuel after ethanol, constitutes 25% of the total biofuel production (Kohler, 2019). The potential raw materials for biodiesel production include edible and nonedible vegetable oils, animal fats, recycled or waste cooking oils, and other saturated and unsaturated fatty acids, of which edible oils are regarded as the basis for the first-generation biodiesel (Hirani, Nasir, Muhammad, Saikat, & Kumar, 2018). The most common and industrial method for converting oil to biodiesel is a chemical reaction known as transesterification in which triglycerides (oils) are reacted with an alcohol (typically methanol) in the presence of a catalyst to produce fatty acid methyl esters (biodiesel) and a high-value coproduct, glycerol (Salvi & Panwar, 2012).

Globally, more than 350 oil-bearing crops have been identified as potential feedstocks for biodiesel production (Hirani et al., 2018). These crops differ from each other in a number of respects, including productivity (i.e., biomass or biodiesel yield per hectare), availability, perishability, production cost, oil composition, etc. (Salvi & Panwar, 2012). Based on the biodiesel yield per hectare as an important criterion for the selection of biodiesel feedstocks, a number of crops with high potential for first-generation biodiesel production are shown in Fig. 1.2. In 2017, about 31% of global biodiesel production was attributed to palm oil, 27% to soybean oil, 20% to rapeseed oil and the remainder to other oils (mainly waste cooking oil and animal fat) (Rezania et al., 2019). The two main advantages that have made these oils the dominant feedstocks for large-scale commercial production of biodiesel are their relatively low cost and readily availability in large quantities (Parcheta et al., 2017).

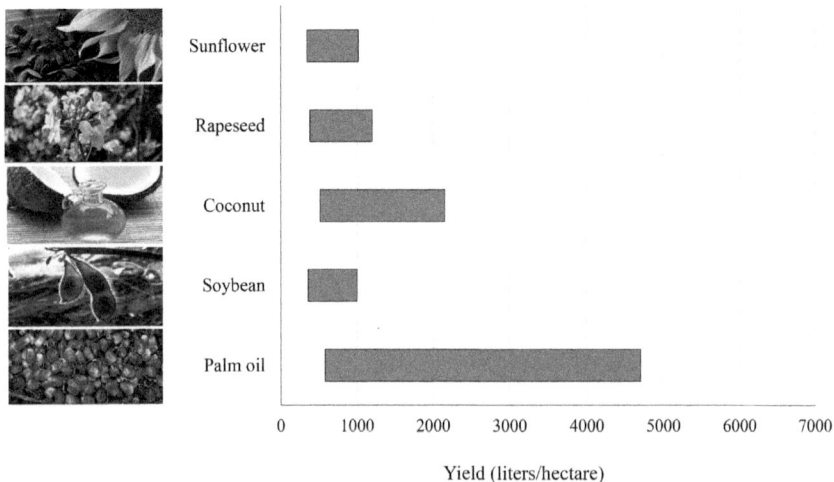

Figure 1.2 *Range of biodiesel yields from various crops.* Based on data from Naylor et al. (2007), Johnston et al. (2009), Leal et al. (2013).

1.3 Second-generation biofuels

The usage of edible crops for biofuel production is in conflict with food production for human consumption, which places unprecedented pressure on global food supply and leads to growing concerns about food security and the viability of first-generation biofuels as an alternative fuel to meet the increasing demand for traditional fossil fuels. To overcome the limitations of first-generation biofuels, second-generation biofuels, also referred to as advanced biofuels, have been developed. They are produced from nonedible feedstocks, thereby reducing dependence on food crops and extending the amount of biofuel that can be produced globally. There are various types of feedstocks for second-generation biofuels which can be broadly categorized into two main groups: lignocellulose feedstocks and nonedible oil feedstocks.

1.3.1 Lignocellulosic feedstocks

Lignocellulosic biomass, as the most abundantly available biomass feedstock throughout the world, can be used for biofuel production without or with very limited negative effects on food security. Lignocellulose is primarily composed of cellulose, hemicellulose, and lignin. Cellulose is a long and stable polymer with repeating units of glucose. Hemicellulose is also a carbohydrate polymer but it can include both five-carbon sugars (e.g., xylose and arabinose) and six-carbon sugars (e.g., glucose and mannose), and has a branched structure with little strength (and little resistance to hydrolysis). Lignin is an amorphous polymer that acts as a reinforcing agent binding all other parts together (Vohra et al., 2014). Due to the recalcitrant nature and rigid structure of lignocellulosic biomass, the processes for converting it into biofuels are more complicated than those used for first-generation biomass feedstocks.

Broadly speaking, there are two basic conversion routes for biofuel production from lignocellulosic biomass, namely biochemical and thermochemical. The biochemical route is similar to the conversion pathway for first-generation biofuels but requires pretreatment. The treatment process that can be physical, chemical, biological, or a combination of them breaks down the biomass structure and makes the cellulose and hemicellulose more accessible for subsequent hydrolysis commonly performed by dilute acid, concentrated acid, and enzymatic hydrolysis (Vohra et al., 2014). The obtained simple sugars after hydrolysis are fermented to ethanol, and the remaining material including mainly lignin can be combusted to provide heat and electricity (Zabed et al., 2017). The thermochemical route involves heating biomass at moderate to high temperatures with limited or no oxygen to change its chemical structures, which results in the production of bio-char, bio-oil, or syngas (Yue et al., 2014).

In general, lignocellulosic biomass for biofuel can be divided into two main categories: organic residues/wastes and dedicated energy crops. The first category includes already available biomass materials that can be found in various sectors such as agriculture and manufacturing, while the second category refers to biomass crops that are specifically grown and utilized for biofuel purposes.

1.3.1.1 Organic residues/wastes

It is estimated that the technical potential of organic residues and wastes worldwide is about 80 exajoules (EJ) per year (Blaschek, Thaddeus, & Scheffran, 2010), corresponding to around 14% of the world primary energy demand and 67% of the world energy demand in the transport sector in 2017, although more research is required to determine how much of this potential can be realized in an environmentally, socially, and economically sustainable way. Depending on where in the value chain (including production, processing and consumption) they arise, residues and wastes fall into three groups: primary, secondary, and tertiary residues.

Primary residues like corn stover and treetops are left in the field and forest during the production and initial processing of crops and forest products (Faaij 2006). Agricultural residues (e.g., wheat straw, rice husk, cotton stalk, and corn leaves) and forest residues (e.g., stems, stumps, barks, branches, twigs, and needles) are two main sources of primary residues, whose annual productions in the world are estimated to be about 5.1 billion and 501 million dry tons, respectively (Ho et al., 2014), representing a tremendous potential for application in the biofuel industry.

Secondary residues are by-products produced during the processing of biomass materials and food crops in associated industries such as food, wood, biofuel. They include crop processing residues (e.g., nutshells, cocoa husk, oil-seed press cakes, sugar cane bagasse, and fruit bunches) and wood processing residues (e.g., sawdust, off-cuts, discarded logs, and bark).

Tertiary residues refer to postconsumer residues obtained after a biomass-derived commodity has served its intended purpose and is to be disposed of. The predominant source of tertiary residues is municipal solid waste, of which a high proportion are organic components standing as suitable feedstocks for biofuel production due to their high carbohydrates, negative price, and availability in huge quantity (Mahmoodi, Keikhosro, & Taherzadeh, 2018).

1.3.1.2 Dedicated lignocellulosic crops

During the transition toward the second-generation biofuel, energy crops that are grown for energy instead of food show great promise as an alternative to food crops thanks to their advantages such as reducing competition with food crops, high biomass yield, low nutrient or fertilizer requirements, ease of cultivation and harvesting, round the year availability, and ability to grow in a broad range of soil types and to survive under extreme environmental conditions (Nanda, Rachita, Prakash, Ajay, & Kozinski, 2018; Srirangan et al., 2012). Dedicated lignocellulosic crops are generally categorized as either woody or herbaceous crops.

Woody crops are fast-growing, short-rotation woody plant species with typical production cycles of 5–8 years. Among the species, poplar (Populus species), illows (Salix species), silver maple (*Acer saccharinum*), black locust (*Robinia pseudoacacia*), eucalyptus (Eucalyptus species), sweetgum (*Liquidambar styraciflua*), loblolly pine (*Pinus taeda*), and sycamore (*Plantanus occidentalis*) are perceived to be the most potential ones (Hinchee et al., 2011). Herbaceous crops are nonwoody, perennial crops that are harvested annually at the end of the growing season after they reach maturity and full productivity (usually taking 2–3 years). There are several examples of herbaceous crops being investigated for energy applications, including miscanthus (Miscanthus species), switchgrass (*Panicum virgatum*), bamboo (Bambusoideae), giant reed (*Arundo donax*), reed canary grass (*Phalaris arundinacea*), tall fescue (Festuca arundinacea), alfalfa (Medicago sativa), and big bluestem (*Andropogon gerardii*) (Hassan, Gwilym, & Jaiswal, 2019; Hughes et al., 2014).

1.3.2 Nonedible oil feedstocks

The current biodiesel production in the world is dominated by first-generation biodiesel obtained from edible oils. However, the relatively high cost of edible oils, the growing criticism over the sustainability of first-generation biodiesel and their competition with the food supply have drawn attention toward nonedible oil-based biodiesel, also known as second-generation biodiesel. In general, the potential nonedible feedstocks for biodiesel production can be categorized as dedicated oil crops (i.e., nonedible vegetable oil), waste cooking oil, and animal fats (Lim et al., 2010).

1.3.2.1 Dedicated oil crops

Dedicated oil crops, also referred to as nonedible oil crops, have two main advantages over traditional edible oil crops for biodiesel production. First, because of their toxic compounds, dedicated oil crops are unsuitable for human consumption, thereby minimizing negative impacts from using edible oil on food security. Second, they are able to grow and even to produce a reasonably high yield under unfavorable conditions for agriculture (e.g., poor soil quality, inadequate rainfall and water shortage, low fertilizer level, and high temperature variation) without requiring intense cultivation practices, making their cultivation more flexible, efficient, and economical compared to edible oil crops (Banković-Ilić Ivana, Olivera, & Vlada, 2012). However, the problem of nonedible oils is that they usually contain large amounts of free fatty acid (FFA), which complicates their conversion to biodiesel and reduces the biodiesel yield (Gui, Lee, & Bhatia, 2008).

There are a large number of oil crops that have been identified as nonedible oil sources for biodiesel production, among which the most promising are jatropha (*Jatropha curcas*), linseed (*Linum usitatissimum*), neem (*Azadirachta indica*), karanja (*Pongamia pinnata*), polanga (*Calophyllum inophyllum*), castor (*Ricinus communis*), mahua (*Madhuca indica*), rubber seed (*Hevea brasiliensis*), cottonseed, jojoba (*Simmondsia chinensis*), moringa (*Moringa oleifera*), and tobacco (*Nicotiana tabacum*) (Ashraful et al., 2014; Mirhashemi, Shayan, Meysam, & Pishvaee, 2018; No, 2011). The oil content of these crops as an important parameter to evaluate the potential of biodiesel feedstocks is indicated in Fig. 1.3.

1.3.2.2 Waste cooking oil and animal fats

Over the last decade, waste cooking oil and animal fats have attracted considerable attention as cost-effective alternatives to vegetable oils for biodiesel production. While the research in this field is still at a preliminary stage, waste cooking oil and animal fats have a notable share in the biodiesel market, accounting for about 10 and 7% of the global biodiesel production in 2017, respectively (Rezania et al., 2019).

The increased biodiesel production from waste cooking oil is attributed to four main reasons. First, waste cooking oil is 2–3 times cheaper than virgin vegetable oils, which can reduce the cost of biodiesel production significantly. Second, since waste cooking oil is no longer usable for human consumption and is banned from being added as an ingredient in animal feed, it can be utilized without affecting food and feed supply. Third, converting waste cooking oil into biodiesel offers a safe, economical and

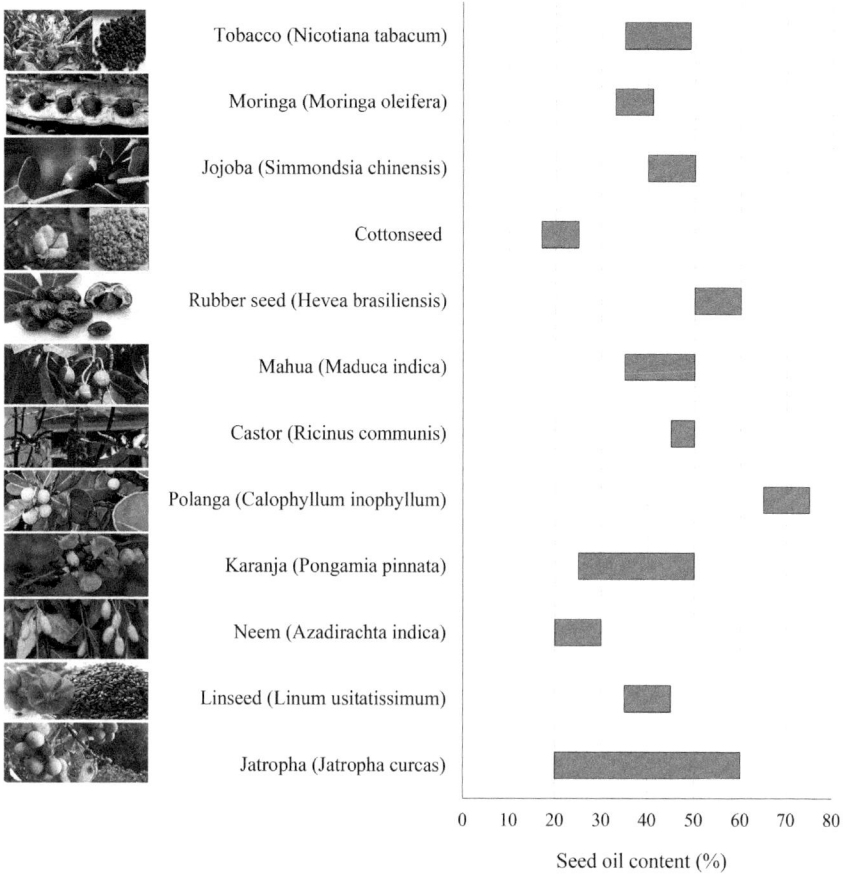

Figure 1.3 *Range of oil content of various oil crops.* Based on data from No (2011), Ashraful et al. (2014), Banković-Ilić et al. (2012).

environmentally friendly alternative way for disposal of waste oil, avoiding environmental problems resulted from improper disposal methods or direct disposal into the environment. Fourth, the amount of waste cooking oil produced by various sources such as restaurants and food industries has been increasing because of the ever-growing human population and food consumption (Gui et al., 2008; Yaakob, Masita, Mohammad, Zahangir, & Sopian, 2013). However, waste cooking oil is contaminated with unwanted contents such as solid impurities, water, and FFA, which significantly affects the quality and yield of biodiesel produced using the conventional transesterification method. Hence, some additional processes must be involved to ensure high yield and high-quality biodiesel meeting international fuel standards (Alptekin, Mustafa, & Sanli, 2014).

Due to their low price and limited market demand, animal fats with low nutritional value for human consumption such as beef tallow, pork lard, poultry fat, and by-products from fishery and meat industries are widely used in some countries as feedstocks for biodiesel production. However, the main drawbacks include limited availability of animal fats and lack of ability to meet the global diesel demand, high FFA content of animal fats and the difficulty of their conversion into biodiesel, and the poor performance of animal fats–derived biodiesel in cold weather (Atabani et al., 2012).

1.4 Third-generation biofuels

Microalgae have emerged as one of the most promising biomass feedstocks for biofuel production. Because of their distinctive and superior properties compared to other feedstocks, microalgae are classified as a new family of biomass resources called third-generation biomass. The advantages of using microalgae as a biofuel feedstock can be summarized as follows:

- Microalgae have a high growth rate which is 100 times faster than terrestrial plants. Generally, they double their biomass within 24 h, and during exponential growth, the biomass doubling time can be as short as 3.5 h (Chisti, 2007; Lam et al., 2012).
- Microalgae are cultivated in open ponds or photobioreactors that can be built even on desert lands. Moreover, some microalgae species are able to grow in brackish, saline and other low-quality water (Mussatto et al., 2010).
- Commonly, the oil content of microalgae species lies in the range of 20%–50% by weight of dry biomass, but it can exceed 80% by optimizing growth conditions. The ability to accumulate a large amount of oil within their cells make microalgae a viable feedstock for biodiesel production, with a projected yield of 47,000–110,000 L/ha (based on an oil-to-biodiesel conversion rate of 80%), which is much more higher than the average biodiesel yield of palm oil (2640 L/ha), coconut (1330 L/ha), rapeseed (790 L/ha), sunflower (680 L/ha), and soybean (678 L/ha) (see Fig. 1.2) (Chisti, 2007).
- Some species contain a high concentration of carbohydrates with no lignin and hemicelluloses content that can be converted into ethanol in a simpler way compared to lignocellulosic biomass requiring chemical and enzymatic pretreatment to break down its complex structure. It is estimated that around 46,760–140,290 L/ha bioethanol can be obtained from microalgae, which is several orders of magnitude greater

than the ethanol yield from sugar and starch crops (see Fig. 1.1) (Mussatto et al., 2010).

- In addition to biofuels, valuable by-products such as proteins, biopolymers, and biomass residue can be made during the production process, which find an application in feed, food, cosmetic and pharmaceutical industries (Ahmad, Mat Yasin, Derek, & Lim, 2011).
- Microalgae cultivation can be tied to concentrated sources of CO_2 such as power plants to achieve high yields of algae and, at the same time, to convert emissions into useful fuel rather than releasing them into the atmosphere (Mohseni and Pishvaee, 2016). Microalgae cultivation can also be coupled to wastewater treatment, which reduces the water and nutrients required for optimal growth and also brings benefits to the treatment process by water contaminants removal (Nodooshan et al., 2018).

1.5 Comparison of three generations of biofuels

This section discusses the advantages and disadvantages of three generations of biofuels under three headlines: (1) land, water, and nutrient requirements, (2) competition with food production, and (3) commercialization and production cost, addressing important issues associated with large-scale biofuel production.

1.5.1 Land, water, and nutrient requirements

First-generation biofuels derived from food crops require extensive arable agricultural lands and consume large amounts of water and other inputs throughout their life cycle. For example, to generate 1 L biodiesel from palm oil, 2 m^2.year land and 5166 L water are needed. These values reach 18 m^2.year and 11,397 L for soybean that has less biodiesel yield per hectare (Mata Teresa, Antonio, & Caetano, 2010; Mekonnen et al., 2011).

Dedicated energy crops have lower land, water, and fertilizer requirements compared to food crops due to their potential to grow on marginal lands, high tolerance to nutrient deprivation, and better water use efficiency (Mehmood et al., 2017). It should be noted that although energy crops are less demanding in terms of water availability, soil quality and fertilization, their productivity can be significantly improved if cultivated on arable lands under sufficient water and fertilizer amounts (Heaton, Tom, & Long, 2004; Smith & Slater, 2010).

Compared to food and energy crops cultivated on arable or marginal lands, no additional land is required for biomass residue production, which avoids competition for land and adverse effects of direct and indirect land-use changes (Carriquiry, Xiaodong, & Govinda, 2011). However, excessive removal of residues can increase fluctuation in soil temperature, exacerbate the risks of soil erosion and global climate change, deplete the soil organic carbon pool and degrade soil quality (Lal, 2005).

Microalgae with a low oil content of 30% need only 0.2 m² (marginal) land annually to produce 1 L biodiesel. However, the life–cycle usage of water, nitrogen, and phosphate are as high as 3726, 0.33, 0.71 kg/L biodiesel, respectively. It should be noted that recycling all harvested water, although it might be impossible or at least very costly in practice, can reduce the water and nutrients consumption by 84% and 55%, respectively (Yang et al., 2011).

1.5.2 Competition with food production

As first-generation biofuels are directly produced from food crops, their production in large quantities leads to the diversion of vast amounts of crops from food markets to bioenergy markets, which drives up food prices and sparks concern about food security. It is estimated that 110 people can be fed by crops used to generate 1 TJ of bioethanol (Rulli, Davide, Andrea, Giulia, & D'Odorico, 2016), which is approximately equivalent to 12,500 gallons (4731.7 L) of bioethanol, showing how important this issue is.

Although second-generation biofuels come from nonfood feedstocks including dedicated energy crops and organic residues, they cannot entirely resolve the food versus fuel conflict. Dedicated energy crops require less amount of inputs (e.g., water, fertilizer, and land), but the commercial-scale production of energy crops creates additional demand for these inputs and gives rise to an increase in their prices. Moreover, the widespread adoption of energy crops results in the reallocation of water, fertilizer, labor, equipment, capital, and other inputs from agricultural to biofuel production, which negatively influences food production and prices. The substantial rise in the demand for residues can also be responsible for negative competition between fuels and food. For example, more demand for corn stover means more land devoted to corn production, so there would be less land available for cultivating other crops, which indirectly contributes to higher prices for those crops (Thompson et al., 2013).

Microalgae do not need a large area of land for growth and can be cultivated on marginal lands without competing with other crops for arable lands. However, due to high water and nutrients requirements of microalgae cultivation, large-scale microalgae production might compete with food production for water and fertilizers, which could put upward pressure on water resources and food prices particularly in countries facing water and food shortage.

1.5.3 Commercialization and production cost

First-generation biofuels have been produced on a large scale for many years around the world. The processes and technologies adopted for converting food crops into biofuels have reached a good level of maturity and high biomass-to-biofuel conversion rate, making first-generation biofuels competitive with their fossil fuel counterparts. Although the production costs of these biofuels may vary depending on geographic region and crop type, they are close to the prices of fossil-based transport fuels, particularly in the cases of US corn ethanol and Brazilian sugarcane ethanol, the world's leading sources of biofuel. Historically, the lowest ethanol production costs have been reported by Brazil, but with the continuously growing ethanol production capacity in the United States combined with multiple crises experienced by the Brazilian ethanol industry, both countries currently have similar production costs (Oliveira, Rochedo, Bhardwaj, Worrell, & Alexandre, 2019).

Although considerable progress has been made in improving their economic and technical performance, second- and third-generation biofuels still face three main barriers to the commercialization:

- High productions cost: The cost estimates of second- and third-generation biofuel production vary widely among different studies, which can be explained by the fact that they are obtained from techno–economic assessments rather than actual data from biorefineries operating at commercial-scale. On average, the production cost of cellulosic ethanol as a superior second-generation biofuel is found to be at least twice the price of gasoline, while microalgae biodiesel is seven times more expensive than diesel (Carriquiry et al., 2011). These differences can explain why high production cost is identified as one of the major barriers confronting the achievement of commercial viability.
- Immature technology: The immediate factor hindering the emergence of an industry transforming lignocellulosic and microalgae into biofuels at a cost-competitive with fossil fuels is the lake of mature and robust technology. The current technologies used for this purpose are relatively young, with low conversion efficiency and high production cost. At present, the main technological challenge of second-generation biofuel production is related to the pretreatment process breaking down

First-generation biofuels		
• Competition with food crops • Need arable land • Low land use efficiency • High water and fertilizer requirement • Low capital and production cost • Experience with commercial production/well-established technologies • High price of feedstocks (food crops)	**Second-generation biofuels**	
	• No food vs. fuel conflict • Marginal land can be used • Low land use efficiency for energy crops • Low water and land requirement • Not commercially available yet/immature technologies • Moderate capital and production cost • Low price of residues but high cost of collection • Reduce waste oils that need to treated	**Third-generation biofuels**
		• No food vs. fuel conflict • High water requirement • Marginal land can be used • High land use efficiency • Not commercially available yet/immature technologies • High capita and production cost • Ability to grow in brackish water • Provide a variety of valuable by-products • Wastewater treatment • CO_2 fixation

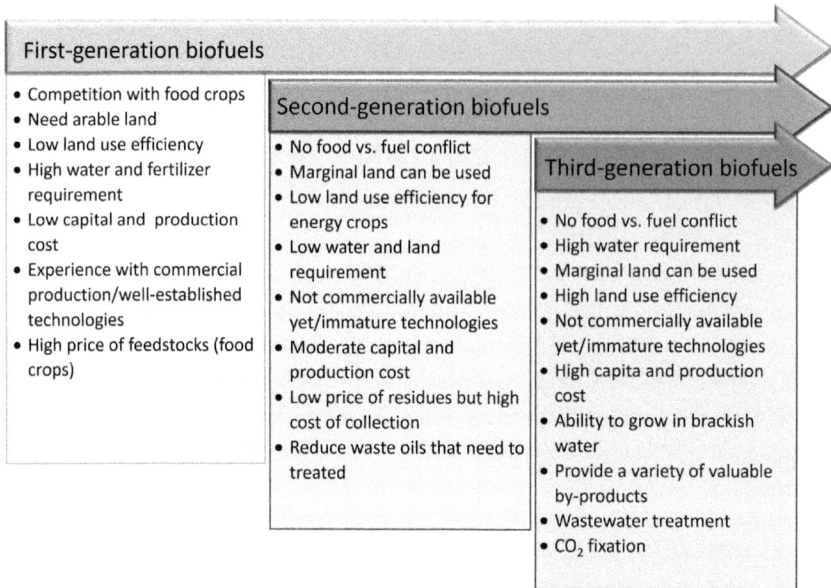

Figure 1.4 *The advantages and disadvantages of three generations of biofuels.*

lignocellulosic biomass into easily hydrolysable components. In the production chain of microalgae–based biofuels, microalgae cultivation, harvesting and drying processes need more technological and cost reduction efforts to accelerate the transition toward the commercial feasibility of third-generation biofuels (Mohseni, Mir, & Sahebi, 2016b).

The advantages and disadvantages of three generations of biofuels are summarized in Fig. 1.4.

1.6 Conclusions

Today's biofuels come primarily from food crops, which has raised serious concerns over food security. However, not only is there no sign of a decline in biofuel production from food crops, but it will probably keep growing due to the lack of a cost–competitive alternative. On the other hand, with increasing global total fuel demand, technological breakthroughs in the biofuel industry, and government interventions in terms of mandates, subsidies, and differential taxation systems, nonfood crops derived biofuels are projected to progressively penetrate the liquid fuel markets in the coming years. Therefore, a diversified portfolio of biomass feedstocks encompassing both food and nonfood crops is most likely to be used for biofuel production in the next one to two decades.

References

Ahmad, A. L., Mat Yasin, N. H., Derek, C. J. C., & Lim, J. K. (2011). Microalgae as a sustainable energy source for biodiesel production: A review. *Renewable and Sustainable Energy Reviews, 15*(1), 584–593.

Alptekin, E., Canakci, M., & Huseyin, S. (2014). Biodiesel production from vegetable oil and waste animal fats in a pilot plant. *Waste Management, 34*(11), 2146–2154.

Anderson, E., Arundale, R., Maughan, M., Oladeinde, A., Wycislo, A., & Voigt, T. (2011). Growth and agronomy of Miscanthus x giganteus for biomass production. *Biofuels, 2*(1), 71–87.

Ashraful, A. M., Masjuki, H. H., Kalam, MdA., Rizwanul Fattah, I. M., Imtenan, S., Shahir, S. A., & Mobarak, H. M. (2014). Production and comparison of fuel properties, engine performance, and emission characteristics of biodiesel from various non-edible vegetable oils: A review. *Energy Conversion and Management, 80*, 202–228.

Atabani, A. E., Silitonga, A. S., Badruddin, I. A., Mahlia, T. M. I., Masjuki, H. H., & Mekhilef, S. (2012). A comprehensive review on biodiesel as an alternative energy resource and its characteristics. *Renewable and Sustainable Energy Reviews, 16*(4), 2070–2093.

Banković-Ilić Ivana, B., Stamenković, O. S., & Veljković, V. B. (2012). Biodiesel production from non-edible plant oils. *Renewable and Sustainable Energy Reviews, 16*(6), 3621–3647.

Blaschek, H. P., Ezeji, T. C., & Scheffran, J. (2010). Biofuels from agricultural wastes and byproducts: An introduction. In *Biofuels from agricultural wastes and byproducts* Wiley, Ames. pp. 3-10.

Bušić, A., Marđetko, N., Kundas, S., Morzak, G., Belskaya, H., Ivančić Šantek, M., Komes, D., Novak, S., & Šantek, B. (2018). Bioethanol production from renewable raw materials and its separation and purification: A review. *Food Technology and Biotechnology, 56*(3), 289.

Carriquiry, M. A., Du, X., & Timilsina, G. R. (2011). Second generation biofuels: Economics and policies. *Energy Policy, 39*(7), 4222–4234.

Chisti, Y. (2007). Biodiesel from microalgae. *Biotechnology advances, 25*(3), 294–306.

EC, *2030 climate and energy framework*. European Commission. 2018. Available from https://ec.europa.eu/clima/policies/strategies/2030_en.

Faaij, A. P. C. (2006). Bio-energy in Europe: Changing technology choices. *Energy Policy, 34*(3), 322–342.

Friesen, P. C., Peixoto, M. M., Busch, F. A., Johnson, D. C., & Sage, R. F. (2014). Chilling and frost tolerance in Miscanthus and Saccharum genotypes bred for cool temperate climates. *Journal of Experimental Botany, 65*(13), 3749–3758.

Goldemberg, José, & Patricia, Guardabassi (2010). The potential for first-generation ethanol production from sugarcane. *Biofuels, Bioproducts and Biorefining: Innovation for a Sustainable Economy, 4*(1), 17–24.

Gui, M. M., Lee, K. T., & Bhatia, S. (2008). Feasibility of edible oil vs. non-edible oil vs. waste edible oil as biodiesel feedstock. *Energy, 33*(11), 1646–1653.

Guldhe, A., Singh, B, Renuka, N., Singh, P., Misra, R., & Faizal, B. (2017). Bioenergy: A sustainable approach for cleaner environment. In *Phytoremediation potential of bioenergy plants* (pp. 47–62). Springer.

Hassan, S. S., Williams, G. A., & Jaiswal, A. K. (2019). Moving towards the second generation of lignocellulosic biorefineries in the EU: Drivers, challenges, and opportunities. *Renewable and Sustainable Energy Reviews, 101*, 590–599.

Heaton, E., Voigt, T., & Long, S. P. (2004). A quantitative review comparing the yields of two candidate C4 perennial biomass crops in relation to nitrogen, temperature and water. *Biomass and Bioenergy, 27*(1), 21–30.

Hinchee, M., Rottmann, W., Mullinax, L., Zhang, C., Chang, S., Cunningham, M., Pearson, L., & Narender, N. (2011). Short-rotation woody crops for bioenergy and biofuels applications. *Biofuels*, 139–156. Springer.

Hirani, A. H., Javed, N., Asif, M., Basu, S. K., & Ashwani, K. (2018). A review on first-and second-generation biofuel productions. In *Biofuels: Greenhouse Gas Mitigation and Global Warming* (pp. 141–154). Springer.

Ho, D. P., Ngo, H. H., & Guo, W. (2014). A mini review on renewable sources for biofuel. *Bioresource Technology, 169*, 742–749.

Hughes, S. R., & Nasib, Q. (2014). Biomass for Biorefining: Resources, Allocation, Utilization, and Policies. In *Biorefineries* (pp. 37–58). Elsevier.

IRENA. (2014). *REmap 2030: A Renewable Energy Roadmap*. Abu Dhabi: International Renewable Energy Agency.

Johnston, M., Foley, J. A., Holloway, T., Kucharik, C., & Chad, M. (2009). Resetting global expectations from agricultural biofuels. *Environmental Research Letters, 4*(1), 014004.

Kamani, M. H., Eş, I., Lorenzo, J. M., Remize, F., Rosselló-Soto, E., Barba, F. J., & Khaneghah, A. M. (2019). Advances in plant materials, food by-products, and algae conversion into biofuels: Use of environmentally friendly technologies. *Green Chemistry, 21*(12), 3213–3231.

Kohler, Marcel (2019). Economic Assessment of Ethanol Production. In *Ethanol* (pp. 505–521). Elsevier.

Lal, R. (2005). World crop residues production and implications of its use as a biofuel. *Environment International, 31*(4), 575–584.

Lam, M. K., & Lee, K. T. (2012). Microalgae biofuels: A critical review of issues, problems and the way forward. *Biotechnology Advances, 30*(3), 673–690.

Langholtz, M., Webb, E., Preston, B. L., Turhollow, A., Breuer, N., Eaton, L., King, A. W., Sokhansanj, S., Nair, S. S., & Downing, M. (2014). Climate risk management for the US cellulosic biofuels supply chain. *Climate Risk Management, 3*, 96–115.

Leal, M. R. L.V., Luiz, A., Horta, Nogueira, & Cortez, L. A. B. (2013). Land demand for ethanol production. *Applied Energy, 102*, 266–271.

Lim, S., & Lee, K. T. (2010). Recent trends, opportunities and challenges of biodiesel in Malaysia: An overview. *Renewable and Sustainable Energy Reviews, 14*(3), 938–954.

Mahmoodi, P., Karimi, K., & Taherzadeh, M. J. (2018). Efficient conversion of municipal solid waste to biofuel by simultaneous dilute-acid hydrolysis of starch and pretreatment of lignocelluloses. *Energy Conversion and Management, 166*, 569–578.

Mata, T. M., Martins, A. A., & Caetano, N. S. (2010). Microalgae for biodiesel production and other applications: A review. *Renewable and Sustainable Energy Reviews, 14*(1), 217–232.

Mehmood, M. A., Ibrahim, M., Rashid, U., Nawaz, M., Ali, S., Hussain, A., & Gull, M. (2017). Biomass production for bioenergy using marginal lands. *Sustainable Production and Consumption, 9*, 3–21.

Mekonnen, M. M., & Hoekstra, A. Y. (2011). The green, blue and grey water footprint of crops and derived crop products. *Hydrology & Earth System Sciences Discussions, 8*(1.), .

Mirhashemi, M. S., Mohseni, S., Hasanzadeh, M., & Pishvaee, M. S. (2018). *Moringa oleifera* biomass-to-biodiesel supply chain design: An opportunity to combat desertification in Iran. *Journal of Cleaner Production, 203*, 313–327.

Mohanty, S. K., & Swain, M. R. (2019). Bioethanol production from corn and wheat: Food, fuel, and future. In *Bioethanol production from food crops* (pp. 45–59). Elsevier.

Mohseni, S., & Pishvaee, M. S. (2016). A robust programming approach towards design and optimization of microalgae-based biofuel supply chain. *Computers & Industrial Engineering, 100*, 58–71.

Mohseni, S., Pishvaee, M. S., & Sahebi, H. (2016). Robust design and planning of microalgae biomass-to-biodiesel supply chain: A case study in Iran. *Energy, 111*, 736–755.

Mukhtar, K., Asgher, M., Afghan, S., Hussain, K., & Zia-ul-Hussnain, S. (2010). Comparative study on two commercial strains of *Saccharomyces cerevisiae* for optimum ethanol production on industrial scale. *BioMed Research International, 2010*, 1–5. doi: 10.1155/2010/419586.

Mussatto Solange, I., Dragone, G., Guimarães, P. M. R., Silva, J. P. A., Carneiro, L. M., Roberto, I. C., Vicente, A., Domingues, L., & Teixeira, J. A. (2010). Technological trends, global market, and challenges of bio-ethanol production. *Biotechnology Advances*, *28*(6), 817–830.

Nanda, Sonil, Dalai, A. K., & Kozinski, J. A. (2016). Supercritical water gasification of timothy grass as an energy crop in the presence of alkali carbonate and hydroxide catalysts. *Biomass and Bioenergy*, *95*, 378–387.

Nanda, S., Rana, R., Sarangi, P. K., Dalai, A. K., & Kozinski, J. A. (2018). A broad introduction to first-, second-, and third-generation biofuels. In *Recent Advancements in Biofuels and Bioenergy Utilization* (pp. 1–25). Springer.

Naylor, R. L., Liska, A. J., Burke, M. B., Falcon, W. P., Gaskell, J. C., Rozelle, S. D., & Cassman, K. G. (2007). The ripple effect: Biofuels, food security, and the environment. *Environment: Science and Policy for Sustainable Development*, *49*(9), 30–43.

No, S.-Y. (2011). Inedible vegetable oils and their derivatives for alternative diesel fuels in CI engines: A review. *Renewable and Sustainable Energy Reviews*, *15*(1), 131–149.

Nodooshan, K. G., Moraga, R. J., Chen, S.-J. G., Nguyen, C., Wang, Z., & Shayan, M. (2018). Environmental and economic optimization of algal biofuel supply chain with multiple technological pathways. *Industrial & Engineering Chemistry Research*, *57*(20), 6910–6925.

Oliveira, C. C. N., Rochedo, P. R. R., Bhardwaj, R., Worrell, E., & Alexandre, S. (2019). Bioethylene from sugarcane as a competitiveness strategy for the Brazilian chemical industry. *Biofuels, Bioproducts and Biorefining*, *14*(2), 286–300.

Pandey, V. C., Bajpai, O., & Singh, N. (2016). Energy crops in sustainable phytoremediation. *Renewable and Sustainable Energy Reviews*, *54*, 58–73.

Parcheta, P., & Datta, J. (2017). Environmental impact and industrial development of biorenewable resources for polyurethanes. *Critical Reviews in Environmental Science and Technology*, *47*(20), 1986–2016.

Rajagopal, D., Sexton, S. E., Roland-Holst, D., & Zilberman, D. (2007). Challenge of biofuel: Filling the tank without emptying the stomach? *Environmental Research Letters*, *2*(4), 044004.

Rezania, S., Oryani, B., Park, J., Hashemi, B., Yadav, K. K., Kwon, E. E., Hur, J., & Cho, J. (2019). Review on transesterification of non-edible sources for biodiesel production with a focus on economic aspects, fuel properties and by-product applications. *Energy Conversion and Management*, *201*, 112155.

Maria Cristina, R., Bellomi, Davide, Cazzoli, A., De Carolis, G., & D'Odorico, P. (2016). The water-land-food nexus of first-generation biofuels. *Scientific Reports*, *6*, 22521.

Saladini, F., Patrizi, N., Pulselli, F. M., Marchettini, N., & Bastianoni, S. (2016). Guidelines for emergy evaluation of first, second and third generation biofuels. *Renewable and Sustainable Energy Reviews*, *66*, 221–227.

Salvi, B. L., & Panwar, N. L. (2012). Biodiesel resources and production technologies—A review. *Renewable and Sustainable Energy Reviews*, *16*(6), 3680–3689.

Sanchez, O. J., & Cardona, C. A. (2008). Trends in biotechnological production of fuel ethanol from different feedstocks. *Bioresource Technology*, *99*(13), 5270–5295.

Simionescu, M., Albu, L.-L., Szeles, M. R., & Bilan, Y. (2017). The impact of biofuels utilisation in transport on the sustainable development in the European Union. *Technological and Economic Development of Economy*, *23*(4), 667–686.

Smith, R., & Slater, F. M. (2010). The effects of organic and inorganic fertilizer applications to Miscanthus × giganteus, *Arundo donax* and *Phalaris arundinacea*, when grown as energy crops in Wales, UK. *GCB Bioenergy*, *2*(4), 169–179.

Srirangan, K., Akawi, L., Moo-Young, M., & Chou, C. P. (2012). Towards sustainable production of clean energy carriers from biomass resources. *Applied Energy*, *100*, 172–186.

Thompson, W., & Meyer, S. (2013). Second generation biofuels and food crops: Co-products or competitors? *Global Food Security*, *2*(2), 89–96.

Van der Merwe, A. B., Cheng, H., Görgens, J. F., & Knoetze, J. H. (2013). Comparison of energy efficiency and economics of process designs for biobutanol production from sugarcane molasses. *Fuel, 105*, 451–458.

Vohra, M., Manwar, J., Manmode, R., Padgilwar, S., & Patil, S. (2014). Bioethanol production: Feedstock and current technologies. *Journal of Environmental Chemical Engineering, 2*(1), 573–584.

Yaakob, Z., Mohammad, M., Alherbawi, M., Alam, Z., & Sopian, K. (2013). Overview of the production of biodiesel from waste cooking oil. *Renewable and Sustainable Energy Reviews, 18*, 184–193.

Yang, J., Xu, M., Zhang, X., Hu, Q., Sommerfeld, M., & Chen, Y. (2011). Life-cycle analysis on biodiesel production from microalgae: Water footprint and nutrients balance. *Bioresource Technology, 102*(1), 159–165.

Yue, D., You, F., & Snyder, S. W. (2014). Biomass-to-bioenergy and biofuel supply chain optimization: Overview, key issues and challenges. *Computers & Chemical Engineering, 66*, 36–56.

Zabed, H., Sahu, J. N., Suely, A., Boyce, A. N., & Faruq, G. (2017). Bioethanol production from renewable sources: Current perspectives and technological progress. *Renewable and Sustainable Energy Reviews, 71*, 475–501.

Ziska, L. H., Runion, G. B., Tomecek, M., Prior, S. A., Torbet, H. A., & Sicher, R. (2009). An evaluation of cassava, sweet potato and field corn as potential carbohydrate sources for bioethanol production in Alabama and Maryland. *Biomass and Bioenergy, 33*(11), 1503–1508.

CHAPTER 2

Biofuel supply chain structures and activities

2.1 Introduction

There are various barriers and complexities preventing the commercialization of biofuel production and must be overcome to enable the transition towards the sustainable production of biofuels and bioenergy products (Awudu & Zhang, 2012; Yue, Fengqi, & Snyder, 2014). To ensure the large-scale and cost-competitive production and use of biofuels, it is of extreme importance to develop an efficient supply chain network design model that systematically designs and optimizes the entire biomass-to-biofuel supply chain from biomass supply sites to final markets, which can be successfully used for the commercial-scale implementation of biofuel projects by taking into account the problems and challenges encountered in real-world supply chains.

Supply chain design is one of the most important problems in biomass-to-biofuel supply chain management, which has received considerable attention from scholars, policymakers, and practitioners around the globe (Sharma et al., 2013b). The current literature on the biofuel supply chain often builds on supply chain design and optimization models that lack generality and scalability (De Meyer et al., 2014). Such models have been restricted to supply chain models proposed for a particular type of biofuel (e.g., bioethanol) from a particular type of biomass (e.g., switchgrass) while addressing a specified part of the supply chain in a limited geographic area taking into account operations associated with decisions of one certain planning level, which is only suitable for small-scale biofuel production. Although such models provide a great insight into particular fields, they do not consider adequately all activities of the biomass-to-biofuel supply chain (Yue et al., 2014), leading to the sub-optimal performance of the biofuel supply chain in practice. Therefore, there is a clear need to provide a holistic structure of the biofuel supply chain that helps the decision-maker to identify the main operations for converting biomass into biofuels. This identification is required for the development of a comprehensive biomass-to-biofuel supply chain design model which is able to systematically manage

Biomass to Biofuel Supply Chain Design and Planning under Uncertainty
http://dx.doi.org/10.1016/B978-0-12-820640-9.00002-7
21

and optimize the entire biomass–to–biofuel supply chains from feedstock production to biofuels distribution, from unit operations to the entire value chain, from strategic to operational planning levels across multiple spatial and temporal scales (Yue et al., 2014).

Designing a biomass–to–biofuel supply chain deals with identifying: (1) the location, capacity, and technology of conversion facilities to convert the available biomass to biofuel and byproducts, (2) the location of intermediate facilities such as collection centers or preprocessing facilities, (3) biomass source sites that serve a specific collection center/preprocessing facility, (4) collection center/preprocessing facility that serve a specific conversion facility, and (5) blending cites/distribution centers used by a specific conversion facility. Decisions about the design and structure of the supply chain are associated with the long-term planning level, which are called strategic decisions with a long-standing impact on the supply chain. On the other hand, managing the logistics activities of a biomass supply chain involve decisions of mid- and short-term planning levels related to biomass cultivation, harvesting, collection, preprocessing/pretreatment, storage, and transportation as well as storage, transportation, and blending/distribution of bio-products.

The rest of this chapter is organized as follows. Section 2.2 presents the general structure of the biomass supply chain and discuses the specific logistics activities and operations of the biomass supply chain, including biomass harvesting and collection, preprocessing, storage, and transportation. Section 2.3 provides a comprehensive overview of biomass conversion pathways from a variety of biomass materials into biofuels and bioenergy. It also describes the different conversion technologies of biomass, including thermochemical (i.e., liquefaction, pyrolysis, gasification, combustion, co-firing, and carbonization), biochemical (i.e., anaerobic digestion and fermentation), and chemical processes (i.e., hydrolysis, solvent extraction, and transesterification). Finally, Section 5.5 is devoted to conclusions.

2.2 General structure of the biomass supply chain

A typical biomass supply chain network consists of several entities, which are mainly divided into three segments, including upstream, midstream, and downstream. The upstream entities co-operate with each other to produce and deliver the required biomass feedstocks to the conversion facility. In the conversion facility, referred to as the midstream segment, biomass is converted to bioenergy and by-products through biomass conversion technologies. The downstream segment refers to post conversion processes, including storage, blending, and distribution of final bioproducts to customers.

Various entities have been considered in the biomass supply chain studies, such as biomass source sites, collection centers, storage sites, preprocessing facilities, intermediate processing sites, conversion facilities, blending sites, distribution centers, and demand zones (Sharma et al., 2013a). However, the main entities include biomass source sites, collection centers/preprocessing facilities, integrated biorefineries, blending sites/distribution centers, and demand zones. Fig. 2.1 illustrates the general structure and activities of the biomass supply chains. The biomass is harvested at source sites, collected and/or preprocessed at intermediate facilities, including collection centers or preprocessing facilities, and then transported to biorefineries. At biorefineries, biomass is converted to biofuel and bioproducts through possible conversion pathways, including thermochemical, biochemical, or chemical processes. Final products include biofuel, power and heat, and biochemicals, which are either shipped to blending centers for blending with corresponding fossil fuels (i.e., in the case of bioethanol and biodiesel) or delivered to distribution centers to distribute to customers.

The main logistics activities of the biomass supply chain include biomass cultivation, harvesting and collection, preprocessing, storage, and transportation as well as transportation, storage, and blending/distribution of bioproducts. Generally, the challenging features of biomass materials such as seasonal nature, scattered geographical distribution, low energy content, low bulk density, and uncertainty of biomass supply due to the various sources of uncertainty make its logistics activities complicated and costly. The major logistics activities of the biomass supply chain are described in detail as following.

• Harvesting and Collection

Harvesting and collection activities deal with procuring the required amounts of biomass from source sites. Harvesting is required for terrestrial feedstocks including agriculture-based biomass such as crop residues and energy crops, as well as forest-based biomass such as energy wood. Harvesting of aquatic biomass, which ranges from very small sizes (micro algae and cyanobacteria) to large seaweeds (macroalgae), is more complicated because of their suspension in water culture. Forced flocculation, filtration, and sedimentation are common methods for harvesting microalgae and cyanobacteria. Moreover, the drying process is required in order to achieve high biomass concentrations. On the other hand, the harvesting method of macroalgae is similar to the harvesting of agricultural crops.

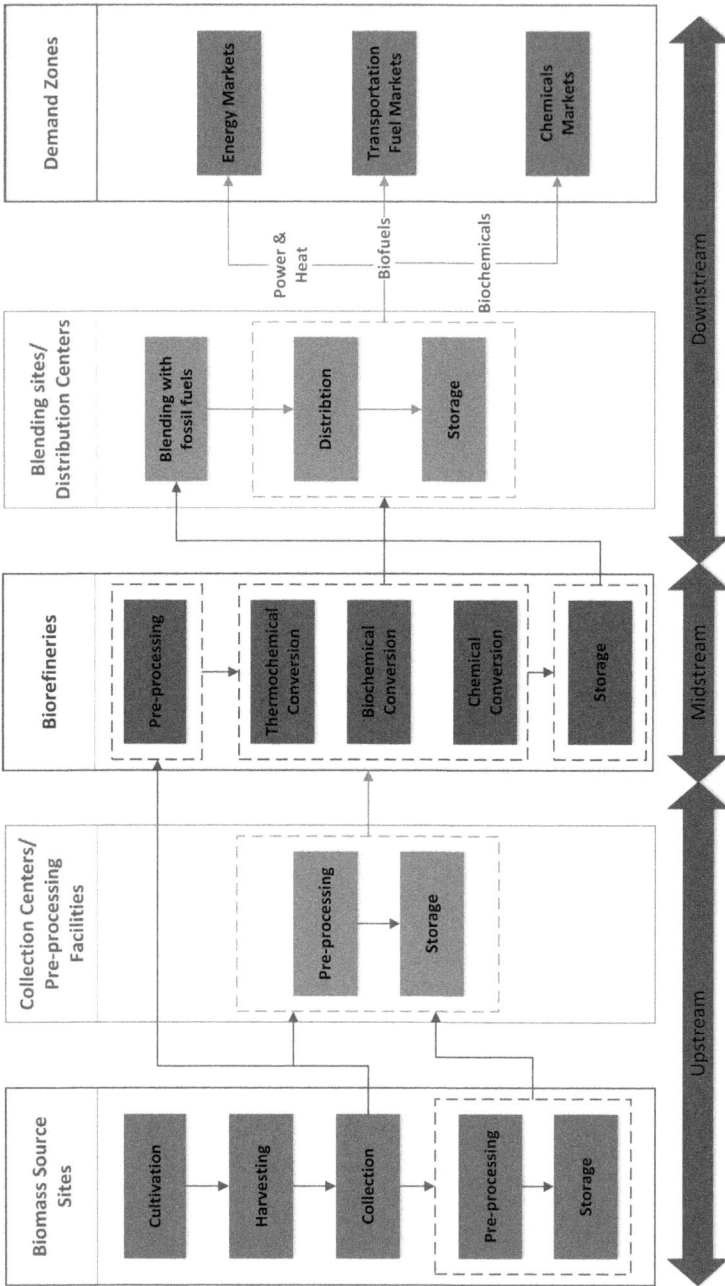

Figure 2.1 A typical biomass supply chain and activities.

Washing and dewatering processes are needed after harvesting of macroalgae. Furthermore, milling is necessary to reduce the large size of macroalgae to a suitable particle size for the conversion process (Lundquist et al., 2010; Yue et al., 2014).

The collection process of agriculture- and forest-based biomass must be performed in a limited time period, because of several factors such as the specific harvesting period of most biomass types, the weather and climate conditions, inaccessibility to forest areas during some months of the year, and the requirement of re-planting the fields. The overlap of this restricted time frame with the collection of food crops may lead to competition between several suppliers in terms of both equipment and labor. As a result, it is necessary to determine the optimal harvest schedule considering available equipment and labor (Malladi & Sowlati, 2018; Rentizelas, Athanasios, & Tatsiopoulos, 2009). Different methods can be applied for collecting biomass based on the feedstock type. For instance, crop residues are usually collected in square bales, round bales, or chopped format, and forest residues are collected in the form of loose or bundled. It should be noted that different collection methods vary in terms of deterioration rates, biomass density, and costs (Malladi & Sowlati, 2018). Consequently, selection of collection methods, routing of collection machinery, and scheduling of equipment and labor within the supply areas are among important logistics decisions of the harvesting and collection processes.

• Storage

The seasonal availability of the most types of biomass, dispersed geographical distribution of biomass and the uncertain nature of biomass supply, on the one hand, and the year-round demand of biomass at conversion facilities, on the other hand, make biomass storage a critical logistics activity of a biomass to biofuel supply chain (Gold & Seuring, 2011; Rentizelas et al., 2009). Storage of biomass can take place at different locations within the supply chain, such as supply areas, biorefineries, and intermediate facilities between the supply areas and biorefineries. Forest residues can be stored at the forest sites for multiple months, which is an appropriate option for reducing the high moisture content of biomass. However, storing agricultural biomass at farms and wood residues at wood processing mills may be constrained by time due to the requirement of preparing land for the subsequent planting season or the limited storage capacity. Accordingly, wood and crop residues are usually not stored in their supply areas for the longterm. On the other hand, the main drawback of storing biomass at biorefineries is the limited storage capacity of these facilities. Storage of biomass at intermediate facilities leads to extra transportation activities to transfer bio-mass from supply areas to intermediate storage sites and from storage facili-ties to biorefineries, which might increases the total logistics costs (Malladi & Sowlati, 2018). Broadly speaking, the storage costs mainly depends on the location and type of storage, the quantity of biomass to be stored, and the storage duration. Different types of storage can be distinguished, such as roof covered or open-air, of which the most suitable must be selected based on the biomass type, and weather and climate conditions. For instance, an indoor storage system is appropriate for wet Japanese weather, while open-air warehouses are suitable to store the biomass with high moisture content. It is worth noting that dry matter losses and quality deterioration during the storage of biomass are side-effects of storing biomass, which can be minimized by the selection of an appropriate storage system. Moreover, long-term storage of large volumes of biomass may increase the risk of fire due to the generated internal heat resulting from the respiration of living cells (Gold & Seuring, 2011).

• Preprocessing

After harvesting, preprocessing operations are typically applied on the raw biomass to reduce its moisture content and particle size, which leads to enhanced feedstock quality, stability, and conversion performance. Each conversion method requires a specified acceptable level of moisture con-tent and particle size for converting biomass, which can be achieved by preprocessing operations (An, Wilbert, & Searcy, 2011; Yue et al., 2014). Moreover, pretreatment processes can be utilized to increase the density of biomass, which eases the handling of raw biomass materials and reduces the cost of transportation (Ba, Christian, & Prodhon, 2016). It is worth not-ing that the requirement of preprocessing depends on the biomass type as well as the corresponding harvesting method. For instance, most agricul-ture-based biomass materials, which are typically collected in bale formats, require grinding before densifying or converting to bioproducts. On the other hand, forest-based biomass residues often need size reduction before they are utilized in the conversion operations, while mill residues consisting shavings and sawdust do not require additional preprocessing due to their suitable form (Malladi & Sowlati, 2018).

Major preprocessing operations include drying, baling, chipping/grind-ing, pelletisation, torrefaction, fast pyrolysis, etc. (Gold & Seuring, 2011; Malladi & Sowlati, 2018; Yue et al., 2014). Drying is utilized to reduce the moisture content of raw biomass and therefore facilitates conversion processes such as gasification and combustion. Drying biomass materials in open-air (i.e., ambient drying) seems as a cost-effective option; however, its effectiveness is dependent on the weather and climate conditions. Baling increases the density of raw biomass and also reduces the risk of deterioration of biomass quality during the storage. This method is mainly adopted for energy crops like miscanthus and forest residues. Chipped biomass can be directly converted chemically or thermally into bioenergy, or it can be processed to pellets form for long-distance transportations. Pelletisation technology is a useful technique for reducing the moisture content as well as increasing the bulk density of biomass (Gold & Seuring, 2011).

Preprocessing of agriculture-based biomass bales is usually performed at the conversion facilities. However, pre-treatment of forest residues must be processed before converting to bio-products. Pre-processing of forest residues can be performed at forest sites, conversion facilities, or intermediate facilities, such that the optimal location must be determined according to the trade-off between transportation and preprocessing costs (Gold & Seuring, 2011).

• Transportation

Low energy density and high moisture content of biomass during long transportation distances lead to high transportation costs and consequently, high logistics costs (Yue et al., 2014). In this respect, preprocessing and densification are useful processes to increase the stability and energy density of biomass for transportation. Generally, several factors affect the transportation costs of biomass, including the mode of transport, distance, the number of loads, the capacity of carriers, the quantity of transported biomass, etc. Train, truck, and barge are feasible transportation modes for biomass transportation. Truck as a widely adopted transportation mode for transporting biomass is cost-effective merely for short distance transportations, while barge and train are cost-effective options for transporting high quantities of biomass over long distances. However, the applicability of these transportation modes is considerably constrained because of limited access of biomass source sites to them (Gold & Seuring, 2011; Malladi & Sowlati, 2018). Tanker truck is the most common transportation mode to transport produced bioethanol and biodiesel from conversion facilities to distribution centers. It is worth noting that pipeline transportation may emerge as an economically feasible alternative in the long term, despite the significant upfront costs and inherent complexities.

2.3 Biomass conversion pathways

Biomass can be converted to several useful end-products, including biofuels, power and heat, or biochemicals. Fig. 2.2 provides a comprehensive schematic overview of biomass conversion pathways from a variety of biomass materials into intermediate and final biofuels and bio-energy products. The conversion of biomass to produce biofuels and bio-energy products can occur using three different categories of conversion technologies, namely thermochemical, biochemical/biological, and chemical processes, as illustrated in Fig. 2.3. These conversion pathways are further described in detail in the following sections.

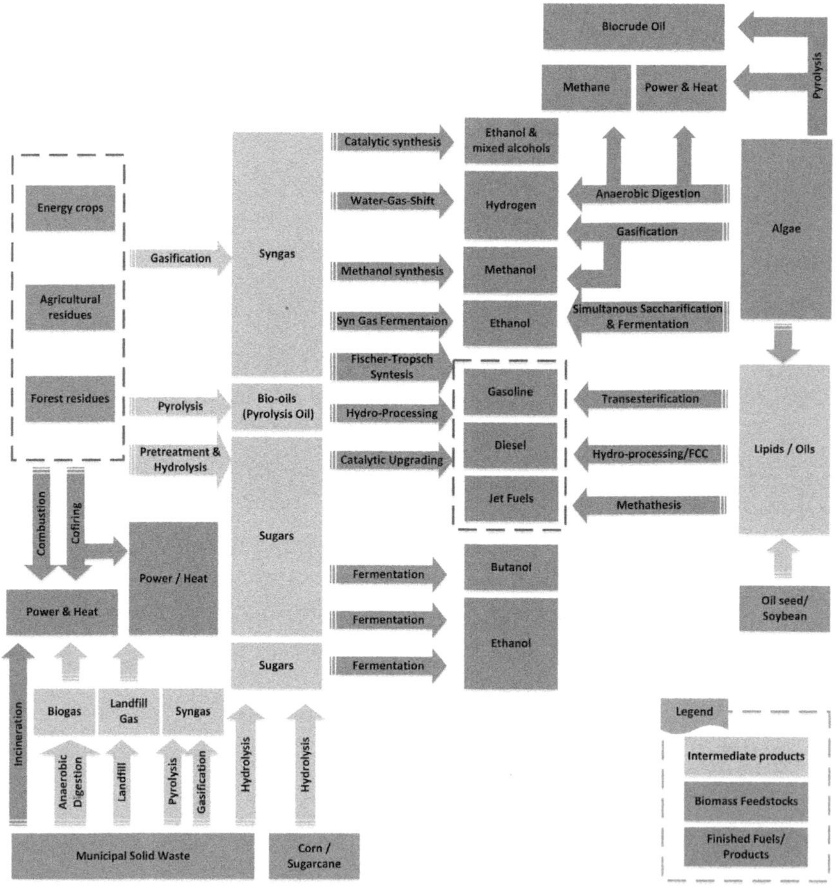

Figure 2.2 *Schematic overview of biomass conversion pathways to intermediate and final products.*

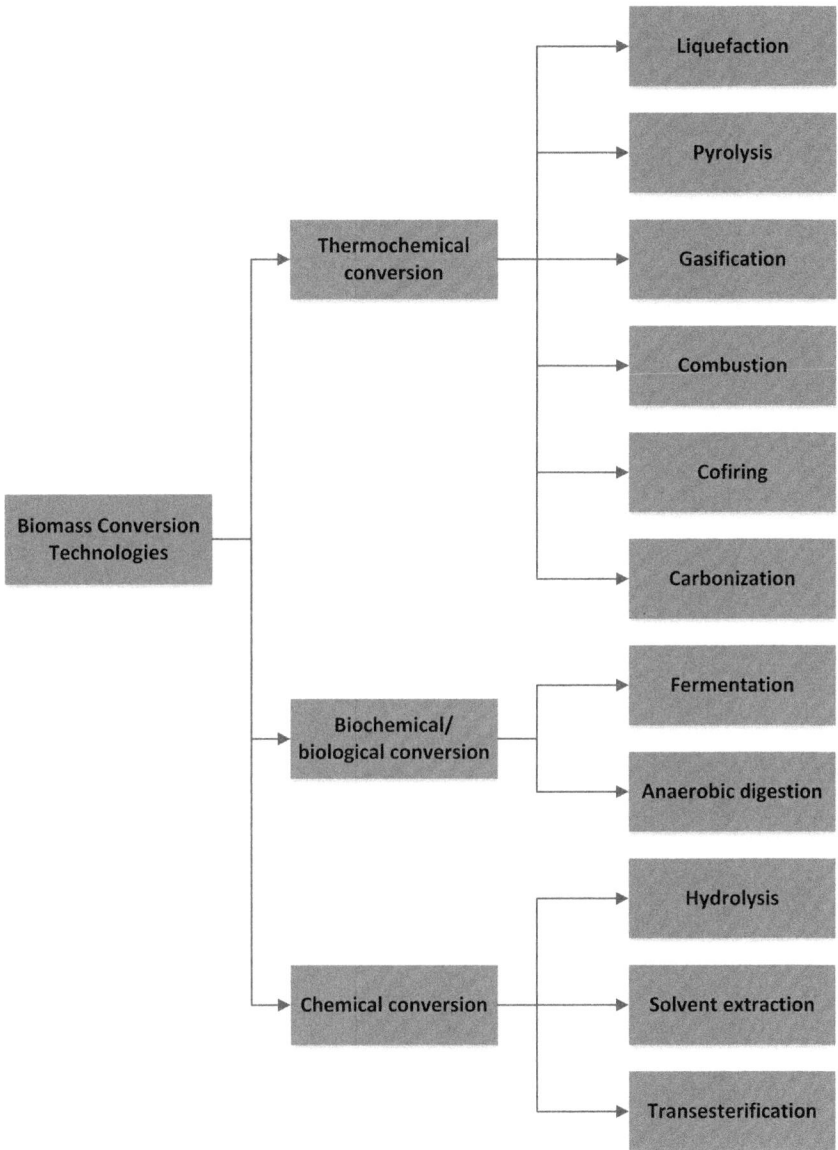

Figure 2.3 *Biomass conversion pathways.*

2.3.1 Biomass thermochemical conversion pathways

The thermochemical processing involves the conversion of biomass to various end products such as biofuels, biochemical, and electricity by treating the biomass under high temperatures in an oxygenic or anoxygenic environment. This conversion process changes the chemical structure and physical characteristics of biomass feedstocks. The thermochemical technologies are broadly classified into the six conversion routes, namely liquefaction, pyrolysis, gasification, combustion, cofiring, and carbonization.

• Liquefaction

The liquefaction process for biomass conversion is commonly performed under a liquid phase at the pressure range of 5–25 MPa with moderate to high temperature range (250–550°C) (Patel, Xiaolei, & Kumar, 2016). Liquefaction pathways are broadly classified into direct liquefaction and hydrothermal liquefaction. Direct liquefaction excludes the gasification step in the conversion process of biomass to liquid fuels (Nanda et al., 2014). The advantage of direct liquefaction in comparison with gasification and pyrolysis pathways is that it does not require drying biomass feedstocks. (Patel et al., 2016). On the other hand, in the hydrothermal liquefaction, solid biomass is directly converted to liquid fuels using water and catalysts (Nanda, Janusz, & Dalai, 2016; Patel et al., 2016). This process reduces the complex structure of biomass and converts biogenic wastes into useful chemicals and heavy oil (Nanda et al. 2014). Hydrothermal liquefaction is an appropriate conversion process for some types of biomass feedstocks such as algal biomass, sewage sludge, municipal waste, and food waste (Adams et al., 2018).

• Pyrolysis

Pyrolysis conversion heats biomass without oxygen, which leads to the thermal degradation of biomass to produce solid (charcoal), liquid (bio-oil), and gaseous products (Naik et al., 2010). Pyrolysis conversion is the initial step of all thermochemical pathways, because in this technology various chemical reactions of biomass are operated in the absence of oxygen to produce three different phases of products (Adams et al., 2018). Based on the operating conditions, such as operating temperature, heating rate, residence time, and particle size of biomass feedstocks, pyrolysis conversion is categorized into six groups: (1) fast pyrolysis, (2) intermediate pyrolysis, (3) slow (or conventional) pyrolysis, (4) flash pyrolysis, (5) vacuum pyrolysis, and (6) ablative pyrolysis (Patel et al., 2016). Fast pyrolysis is a widely adopted technology since the yield of pyrolysis oil is highly enhanced. The obtained pyrolysis oil can be upgraded to hydrocarbon fuels using hydrotreating and hydrocracking processes (Yue et al., 2014). Moderate temperature range (400–550°C), very short residence time (0.5–2 s), and small particle size (<3 mm) are the main characteristics of this conversion route (Patel et al., 2016). Slow pyrolysis occurs in the moderate temperature range (350–750°C), with a low heating rate and long residence time, which is mainly used to produce charcoal (Patel et al., 2016). Flash pyrolysis can be distinguished from the slow pyrolysis by its relatively shorter residence time and higher heating rate and it is often adopted to produce bio-oil that can be merged with char to produce bioslurry (Naik et al., 2010).

- Gasification

In the gasification technology, biomass is reacted with a gasifying agent (e.g., steam, oxygen or air) at moderate to high temperature range (600–1200°C) to produce a combustible gaseous mixture of H_2, N_2, CO, CO_2 and CH_4, also known as synthesis, syngas or producer gas (Naik et al., 2010; Patel et al., 2016), which can be utilized to produce heat, power, and chemical products (Adams et al., 2018; Nanda et al., 2014). Furthermore, the syngas can be employed by the Fischer–Tropsch (F–T) synthesis to generate hydrocarbon liquid fuels (Yue et al., 2014).

- Combustion

Biomass combustion is the process of burning biomass in the presence of oxygen, which is recognized as the simplest thermochemical conversion technology. The combustion of biomass involves sequential heterogeneous and homogeneous reactions that generate heat and power. In the combustion process, the chemical energy of biomass feedstocks, such as agricultural residues, forest residues, and municipal solid wastes, can be released. Although this method is one of the conventional methods of generating heat and or power, its negative environmental impacts resulting from high emissions of CO_2, CO, NO_x, and particulate matter are recognized as a major barrier to its application (Patel et al., 2016; Srirangan et al., 2012).

- Cofiring

Biomass cofiring refers to the concurrent blending and combustion of biomass materials with other fuels such as natural gas and coal within a boiler, which reduce the use of fossil fuels for energy generation and emissions without significantly increasing costs and infrastructure investments. In the case of solid biomass feedstocks such as pellets and wood chips, cofiring means the combustion of biomass in coal-fired power plants, whereas for gasified biomass, the concurrent firing of biomass occurs with pulverized coal or natural gas. It should be noted that, depending on the adopted technology, methods of cofiring biomass with coal or natural gas are mainly classified into three categories, namely direct cofiring, indirect cofiring, and parallel cofiring. These methods differ in terms of percentage of biomass in the mixture and the design of boiler systems. Direct cofiring as the most common and low-cost technology, means cofiring the biomass with coal within a boiler. Indirect cofiring technology is cofiring biomass in a gas-fired or oil-fired system, while in the parallel cofiring an entirely distinct external boiler is installed to produce steam utilized for generating electricity in the power plant (Agbor, Xiaolei, & Kumar, 2014; Al-Mansour and Zuwala, 2010).

• Carbonization

Based on the desired amount of energy carrier, in the form of gas-liquid or gas, applications of the pyrolysis process can be extended from slow heating to flash pyrolytic conversion pathways. Pyrolytic conversion of biomass under slow heating is called carbonization, in which a carbon-rich solid residue is the main product compared to gas and liquid yields. This technology is usually adopted to produce charcoal, whereas the extra oils and gas can be recovered for further use (Strezov et al., 2007).

2.3.2 Biomass biochemical/biological conversion pathways

Biochemical conversion utilizes enzymes or microorganisms (e.g., bacteria, yeasts, cyanobacteria, algae) to catalyze the conversion of biomass into liquid or gaseous biofuels, value-added products (e.g., carotenoids) and other commercial chemicals (e.g., acetic acid). The two main pathways of biochemical conversion are anaerobic digestion and fermentation, which have been recognized as green technologies for biomass conversion in terms of the carbon cycle (Fiorentino, Maddalena, & Ulgiati, 2017; Gouveia & Passarinho, 2017).

• Fermentation

In the fermentation pathway, an organic substrate of biomass is chemically changed by the action of enzymes, which is secreted by different microorganisms such as yeasts. Fermentation processes can be classified into two main types, namely aerobic and anaerobic, based on the presence or absence of oxygen (Naik et al., 2010). Fermentation is a widely adopted technology in different countries to produce ethanol (C_2H_5OH) on a large-scale from sugar crops such as sugarcane and sugar beet, as well as starch crops such as maize and wheat. Molasses is the most common biomass feedstock for ethanol fermentation because its total weight is composed of about 50% sugar and about 50% organic and inorganic compounds. The fermentation of sugars for ethanol production has been recognized as one of the greenest technology for liquid fuel production since the generated CO_2 can be used in other applications of a biorefinery. In this process, initially, the biomass feedstock is ground down, and the starch is converted to sugars by enzymes, and finally, the sugars are converted to ethanol by yeasts. Hence, the fermentation of starch is more complex than sugar fermentation, because the starch should be converted to sugar first and then to ethanol (Faruq, Tanvir, & Yusof, 2015; McKendry, 2002).

- Anaerobic digestion

The anaerobic digestion is the process of converting organic material to biogas, which is composed of methane, carbon dioxide, water vapor, and minor quantities of hydrogen sulfide and, in some cases, hydrogen (Vertès, Qureshi, Yukawa, & Blaschek, 2011). In this technology, the biomass and other biodegradable matter are broken down by means of microorganisms in the absence of oxygen (Faruq et al., 2015). The biogas can be mainly produced by anaerobic digestion of municipal/industrial waste waters, municipal solid wastes, and agricultural residues (Soetaert & Vandamme, 2009). Obtained methane gas from this technology can be utilized for electricity generation. The general process of anaerobic digestion is consists of four main steps, namely hydrolysis, acidification, acetogenesis, and methanogenesis (Faruq et al., 2015). It is worth noting that, the anaerobic digestion, as an environmentally friendly technology, has gained much attention through the years, because of its application as a waste treatment method in addition to the production of renewable fuels.

2.3.3 Biomass chemical conversion pathways

Chemical conversion routes refer to processes, such as hydrolysis, solvent extraction, and transesterification, which directly convert biomass materials into chemicals by changing the structure of the substrate at moderate pressure and/ or temperature and in the presence of a catalyst. Chemical conversions can be utilized in the biomass pretreatment process, and also in the downstream processing steps to transform the intermediates produced from biological and thermochemical conversions into final products (Fiorentino et al., 2017).

- Hydrolysis

The three main components of biomass, which are processed by hydrolysis conversion involve cellulose, hemicellulose, and lignin. In the hydrolysis process, these complex compounds are broken down into simple sugar molecules, which are then fermented to produce bioethanol. The critical factors influencing the efficiency of hydrolysis include the biomass ratio of surface to volume, acid concentration, process time, and temperature. It should be noted that the surface to volume ratio of biomass determines the extent of glucose yield. Accordingly, smaller particle size leads to more effective hydrolysis in terms of reaction rate, and the higher amount of liquid to solid ratio leads to faster reaction (Faruq et al., 2015; Naik et al. 2010).

- Solvent extraction

The main steps of the solvent extraction process are: (1) extracting oil from oil seeds by using hexane as a solvent, (2) evaporating the employed solvent, (3) distilling oil–hexane mixture (i.e., miscella), and (4) toasting the de-oiled meal. Halogenated solvents, ethanol, and acetone are among other solvents, which can be applied in specific cases. The extraction process is performed by removing the desired substance from the raw material, dissolving the substance in the solvent, and finally recovering it from the solvent. It is worth noting that extraction and separation are essential operations to remove the specified substance from biomass (Naik et al., 2010).

• Transesterification

In the transesterification process, the extracted oil from biological sources such as vegetable oils and animal fats is chemically reacted with an alcohol like methanol or ethanol in the presence of a catalyst, forming a mixture of methyl esters, commonly known as biodiesel, and glycerin. It should be noted that biodiesel as an environmentally friendly fuel is compatible with existing vehicle engines, while direct vegetable oils can be utilized as fuel in the modified vehicle engines (Gaurav, Sivasankari, Kiran, Ninawe, & Selvin, 2017; Vertès, Qureshi, Yukawa, & Blaschek, 2011).

2.4 Conclusions

Most previous studies on biomass supply chain design and optimization build on models that lack generality and scalability. These models have been restricted to supply chains proposed for producing a particular type of biofuel from a particular type of biomass addressing a specific part of the supply chain in a limited geographic area taking into account specified operations associated with decisions of one certain planning level, which is only suitable for small-scale biofuel production. Therefore, it is of extreme importance to develop an efficient supply chain design model that systematically designs and optimizes the entire biomass-to-biofuel supply chain from biomass supply sites to final markets, which can be successfully used for the commercial-scale implementation of biofuel projects by taking into account the problems and challenges encountered in real-world supply chains. In this respect, this chapter presents the general structure of the biomass supply chain and discusses logistics activities and operations of the biomass supply chain such as biomass harvesting, collection, storage, preprocessing, and transportation are described. Moreover, an overview of conversion pathways from a variety of biomass materials into biofuels and bioenergy products is provided, and different conversion technologies, including thermochemical, biochemical, and chemical processes are presented.

References

Adams, P., Bridgwater, T., Lea-Langton, A., Ross, A., & Watson, I. (2018). Biomass conversion technologies. In *Greenhouse Gas Balances of Bioenergy Systems* (pp. 107–139). Elsevier.

Agbor, E., Zhang, X., & Kumar, A. (2014). A review of biomass co-firing in North America. *Renewable and Sustainable Energy Reviews, 40*, 930–943.

Al-Mansour, F., & Zuwala, J. (2010). An evaluation of biomass co-firing in Europe. *Biomass and Bioenergy, 34*(5), 620–629.

An, H., Wilhelm, W. E., & Searcy, S. W. (2011). Biofuel and petroleum-based fuel supply chain research: A literature review. *Biomass and Bioenergy, 35*(9), 3763–3774.

Awudu, I., & Zhang, J. (2012). Uncertainties and sustainability concepts in biofuel supply chain management: A review. *Renewable and Sustainable Energy Reviews, 16*(2), 1359–1368.

Ba, B. H., Prins, C., & Prodhon, C. (2016). Models for optimization and performance evaluation of biomass supply chains: An Operations Research perspective. *Renewable Energy, 87*, 977–989.

De Meyer, A., Cattrysse, D., Rasinmäki, J., & Orshoven, J. V. (2014). Methods to optimise the design and management of biomass-for-bioenergy supply chains: A review. *Renewable and Sustainable Energy Reviews, 31*, 657–670.

Faruq, M., Arfin, T., & Yusof, N. A. (2015). Chemical processes and reaction by-products involved in the biorefinery concept of biofuel production. In *Agricultural Biomass Based Potential Materials* (pp. 471–489). Springer.

Fiorentino, G., Ripa, M., & Ulgiati, S. (2017). Chemicals from biomass: Technological versus environmental feasibility. A review. *Biofuels, Bioproducts and Biorefining, 11*(1), 195–214.

Gaurav, N., Sivasankari, S., Kiran, G. S., Ninawe, A., & Selvin, J. (2017). Utilization of bioresources for sustainable biofuels: A review. *Renewable and Sustainable Energy Reviews, 73*, 205–214.

Gold, S., & Seuring, S. (2011). Supply chain and logistics issues of bio-energy production. *Journal of Cleaner Production, 19*(1), 32–42.

Gouveia, L., & Passarinho, P. C. (2017). Biomass conversion technologies: Biological/biochemical conversion of biomass. In *Biorefineries* (pp. 99–111). Springer.

Lundquist Tryg, J., Woertz, I. C., Quinn, N. W. T., & Benemann, J. R. (2010). A realistic technology and engineering assessment of algae biofuel production. *Energy Biosciences Institute, 1*.

Malladi, K. T., & Sowlati, T. (2018). Biomass logistics: A review of important features, optimization modeling and the new trends. *Renewable and Sustainable Energy Reviews, 94*, 587–599.

Naik, S. N., Goud, V. V., Rout, P. K., & Dalai, A. K. (2010). Production of first and second generation biofuels: A comprehensive review. *Renewable and Sustainable Energy Reviews, 14*(2), 578–597.

Nanda, S., Kozinski, J. A., & Dalai, A. K. (2016). Lignocellulosic biomass: A review of conversion technologies and fuel products. *Current Biochemical Engineering, 3*(1), 24–36.

Nanda, S, Mohammad, J., Reddy, S. N., Kozinski, J. A., & Dalai, A. K. (2014). Pathways of lignocellulosic biomass conversion to renewable fuels. *Biomass Conversion and Biorefinery, 4*(2), 157–191.

Patel, M., Zhang, X., & Kumar, A. (2016). Techno economic and life cycle assessment on lignocellulosic biomass thermochemical conversion technologies: A review. *Renewable and Sustainable Energy Reviews, 53*, 1486–1499.

Peter, M. (2002). Energy production from biomass (part 2): Conversion technologies. *Bioresource Technology, 83*(1), 47–54.

Rentizelas, A. A., Tolis, A. J., & Tatsiopoulos, I. P. (2009). Logistics issues of biomass: The storage problem and the multi-biomass supply chain. *Renewable and Sustainable Energy Reviews, 13*(4), 887–894.

Sharma, B., Ingalls, R. G., Jones, C. L., & Khanchi, A. (2013a). Biomass supply chain design and analysis: Basis, overview, modeling, challenges, and future. *Renewable and Sustainable Energy Reviews*, *24*, 608–627.

Sharma, B., Ingalls, R. G., Jones, C. L., & Khanchi, A. (2013b). Biomass supply chain design and analysis: Basis, overview, modeling, challenges, and future. *Renewable and Sustainable Energy Reviews*, *24*, 608–627.

Soetaert, W., & Vandamme, E. J. (2009). *Biofuels in perspective. Biofuels*. London, UK: John Wiley & Sons, Ltd.

Srirangan, K., Akawi, L., Moo-Young, M., & Chou, C. P. (2012). Towards sustainable production of clean energy carriers from biomass resources. *Applied Energy*, *100*, 172–186.

Strezov, V., Patterson, M., Zymla, V., Fisher, K., Evans, T. J., & Nelson, P. F. (2007). Fundamental aspects of biomass carbonisation. *Journal of Analytical and Applied Pyrolysis*, *79*(1–2), 91–100.

Vertès, A. A., Qureshi, N., Yukawa, H., & Blaschek, H. P. (2011). *Biomass to biofuels: Strategies for global industries*. United Kingdom: John Wiley & Sons.

Yue, D., You, F., & Snyder, S. W. (2014). Biomass-to-bioenergy and biofuel supply chain optimization: Overview, key issues and challenges. *Computers & Chemical Engineering*, *66*, 36–56.

CHAPTER 3

Decision-making levels in biofuel supply chain

3.1 Introduction

Biomass-to-biofuel supply chain design and planning includes a large number of decisions associated with various operations along the chain such as biomass cultivation and harvesting, biomass transportation, biomass to biofuel conversion, and biofuel transportation and distribution to the final markets. Depending on their timeframe, these decisions are categorized into strategic (long-term), tactical (medium-term), and operational (short-term). Strategic decisions refer to those decisions with a long-standing impact on the supply chain that are difficult or even impossible to change or modify in a short period, mainly involving decisions related to the characteristics of supply chain facilities such as the location, capacity, and technology of biorefineries. Tactical decisions such as biomass and biofuel inventory planning are made within the established strategic decisions over a medium-term horizon ranging from a few months to one year. Operational-level decisions that come after the tactical decision-making are weekly, daily, or hourly decisions focusing on routine day-to-day operations, including scheduling of biomass harvest operations and truck routing for biomass transportation from supply areas to biorefineries (De Meyer, Cattrysse, Rasinmäki, & Van Orshoven, 2014).

The supply chain planning matrix has been introduced as an architecture to hierarchically structure all decisions involved in biofuel supply chain planning, enabling the decision maker to decompose the decision problem into smaller subproblems and handle each of them with an appropriate solution methodology (Fleischmann, Meyr, & Wagner, 2005). As shown in Fig. 3.1, this matrix covers the strategic, tactical, and operational decisions across four key supply chain functional processes including biomass supply and preprocessing, biofuel production, biofuel blending and distribution, and biofuel sales. Biomass supply encompasses the operations associated with the upstream segment of the chain from biomass cultivation, harvesting, and storage to the delivery to biorefineries. It also incorporates biomass preprocessing that refers to chemical and mechanical processes adopted to

Biomass to Biofuel Supply Chain Design and Planning under Uncertainty
http://dx.doi.org/10.1016/B978-0-12-820640-9.00003-9

Biomass Supply and Preprocessing	Biofuel Production	Biofuel Blending and Distribution	Biofuel Sale
Long-term planning			
• Selection of biomass types • Selection of biomass cultivation sites • Determining the location, and capacity of storage/preprocessing centers	• Determining the number, location, technology and capacity of biorefineries • Capacity expansion Technology upgrading	· Biofuel distribution network design • Integrated biofuel and petroleum supply chain design	•Biofuel market selection • Designing the international trade network of biomass/biofuels
Medium-term planning			
•Harvest and collection planning • Biomass inventory planning • Biomass Transportation planning	• Biorefinery process synthesis and design • Production planning • Capacity expansion Technology upgrading	· Biofuel distribution planning • Biofuel inventory planning	·Biofuel pricing • Biofuel demand forecasting
		Master Planning	
Short-term planning			
• Scheduling of harvest and collection operations • Scheduling and routing of vehicles for biomass transportation • Scheduling of preprocessing operations	• Scheduling of production units •Allocation of raw materials, labor, and equipment in biorefineries	· Scheduling and routing of vehicles for biofuel transportation • Scheduling of blending tasks	• Biofuel and byproducts demand fulfillment

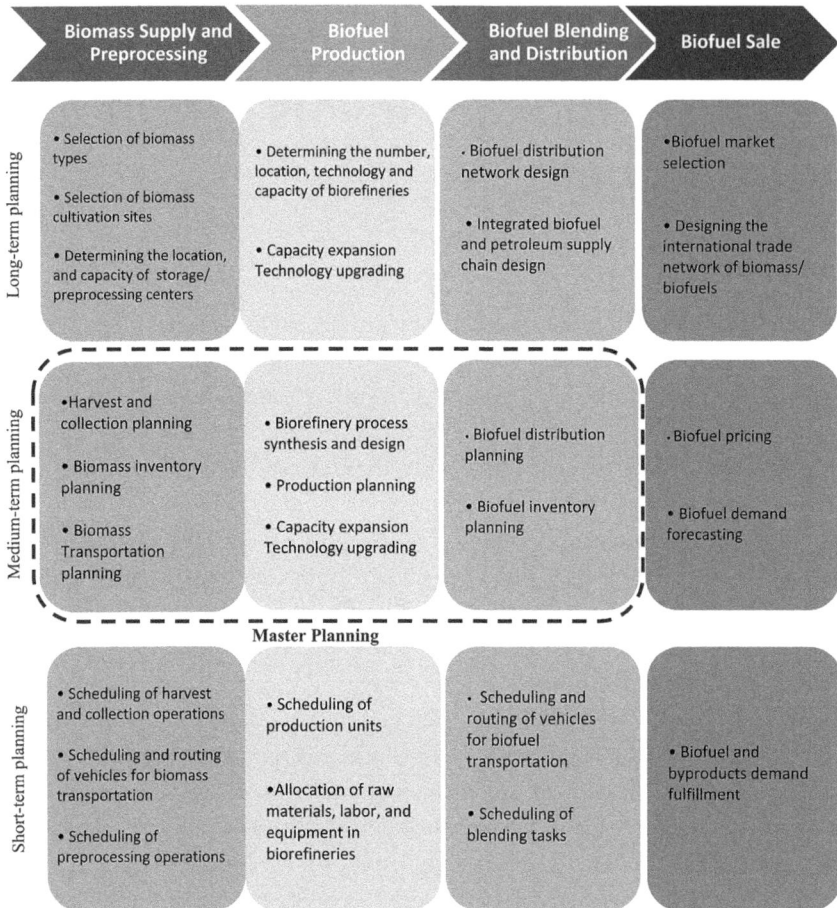

Figure 3.1 *An overview of strategic, tactical, and operational decisions in the biofuel supply chain.*

increase the stability and energy density of biomass with the purpose of facilitating biomass storage, handling, and transportation as well as preparing it for the conversion processes. The constrained capacity of biomass resources is the input of the biofuel production stage, which can involve multiple subprocesses for the conversion of biomass to biofuel. The biofuel blending and distribution process is a layer between production facilities and customers, which can be performed in storage facilities, distribution/blending sites, or petroleum refineries that blend biofuel with fossil fuel counterparts. All of the above supply chain and logistics operations are managed according to demand forecasts and orders of biofuel products coming from the biofuel sales stage.

Locally optimal decisions at each box of the matrix do not necessarily result in optimal performances for the entire biofuel supply chain because of the presence of conflicting objectives and lack of coordination between the various decision levels. To improve the overall performance of the supply chain, there is a consensus in the literature that two or more of these decision boxes must be modeled and optimized together with the help of two common strategies, namely, horizontal and vertical integration. The first integrates the related decisions at the same decision level (i.e., strategic, tactical, or operational), for example, combining biomass type selection with biorefinery location selection, while the second integrates the related decisions from the different levels (i.e., strategic-tactical, tactical-operational, or strategic-tactical-operational), for example, combining biomass transportation planning with truck routing and scheduling (Pishvaee, Farahani, & Dullaert, 2010). Strategic, tactical, and operational decision levels along with their corresponding mathematical optimization models are presented in Chapters 7–9.

The remaining of this chapter is organized as follows: Sections 3.2–3.4 describe the biofuel supply chain decisions that are classified into strategic, tactical, and operational decisions according to the supply chain process and planning horizon, as the two dimensions of the planning matrix. Section 3.5 presents a systematic review of biomass supply chain design and planning literature covering different strategic, tactical, and operational decisions, including identified research gaps. Finally, Section 3.6 provides conclusions and potential future research directions.

3.2 Strategic-level decisions

3.2.1 Selection of biomass type

Selecting the most suitable biomass sources is one of the most important strategic decisions regarding sustainable biofuel production. Broadly speaking, various criteria can be investigated to select among potential biomass alternatives based on their main characteristics such as production cost, yield (i.e., amount of biomass produced per hectare), energy content (i.e., amount of biofuel production per biomass unit), output-to-input energy ratio (i.e., efficient use of light, nutrients, and water), input requirements (i.e., water, energy, fertilizer), ability to grow on degraded or marginal lands, adaptability to different soils and climate conditions, resistance to stress factors such as low and high temperature, nutrient deficiencies, salinity, flooding and drought, and all-year round availability and flexibility in harvest time

(Anderson et al., 2011; Langholtz et al., 2014; Nanda, Dalai, & Kozinski, 2016). On the other hand, food crisis issue is another determining factor in identifying the biomass sources for biofuel production. In this regard, first-generation biofuel production from agricultural crops such as sugarcane and corn that are in competition with the food supply has led to intensify food crisis and increase food prices. Moreover, the high cost of utilizing vegetable oils such as soybean and palm oil that are major feedstocks for biodiesel production has resulted in commercial infeasibility of biodiesel production, as well as intensifying food versus fuel debate (Babazadeh, Razmi, Rabbani, & Pishvaee, 2017). As a result, the focus is now on the utilization of lignocellulosic and low-price nonedible biomass resources for biofuel production (Sharma, Ingalls, Jones, & Khanchi, 2013).

3.2.2 Selection of biomass cultivation sites

In order to determine suitable locations for biomass cultivation, there is a need to evaluate the geographical and climate conditions of the regions where biomass is to be cultivated. In this regard, geographic information system (GIS) can be used as a powerful tool to address the question of where to locate biomass cultivation sites so that their suitability in terms of factors such as land slope, soil PH, annual rainfall, land use, solar radiation, ambient temperature, and distance from water sources is ensured.

Given the growing concerns about the sustainability of biomass production and the environmental and social impacts resulting from their use, it is also of significant importance to consider economic, environmental, and social criteria when determining cultivation locations. Economic aspects that refer to the costs of cultivation and the effect of biomass cultivation on the overall supply chain profitability is typically considered in most of the studies that have investigated the selection of biomass cultivation sites (Bairamzadeh, Pishvaee, & Saidi-Mehrabad, 2016; Gonela, Zhang, Osmani, & Onyeaghala, 2015). The environmental impacts that might occur as a result of large-scale biomass production include but not limited to soil erosion, carbon emissions, changes in carbon and nutrient balance of soil, land-use changes, biodiversity loss, water pollution and eutrophication, water and mineral resource depletion, and fragmentation and deterioration of habitats (Myllyviita, Holma, Antikainen, Lähtinen, & Leskinen, 2012). The social dimension of the sustainability aims to improve the social benefits from biomass production by maximizing its positive impacts such as employment opportunities, income generation, and energy security while minimizing its negative impacts such as food insecurity and increase in food prices due to the diversion of food crops, land, water, and other resources from human needs to biomass production (Ribeiro, 2013).

3.2.3 Selection of facility location, technology, and capacity

Similar to what was said about cultivation site selection analysis, various criteria can be considered to investigate the suitability of potential locations for biorefineries and other supply chain facilities such as storage, preprocessing, and blending sites. For instance, Healey, Irmak, Hubbard, and Lenters (2011) defined a set of environmental and competition criteria in the determination of suitable locations for corn-based ethanol biorefineries. These criteria include corn yields, corn land abundance, corn irrigation requirement, irrigation accessibility and restriction, growing season precipitation, and potential competition between ethanol plants for corn resources. They concluded that factors related to land and water sources play a more significant role in the selection of the biorefinery location than those related to the potential competition for resources. Mohseni, Pishvaee, and Sahebi (2016) categorized the criteria for assessing locations for microalgae production facilities into three groups, namely climate, resources, and infrastructure. Climate criteria that include solar radiation and air temperature restrict the site selection analysis to areas with optimal irradiance and temperature range for the growth of microalgae. Since microalgae are grown in a liquid medium supplied with sufficient amount of water, CO_2, and nutrients, the second group of criteria ensures that microalgae production facilities are located in the vicinity of power plants (i.e., CO_2 sources), water sources, and wastewater treatment plants (i.e., nutrients sources). The infrastructure criteria related to the design and implementation of production facilities include distance to road, land use, and land slope.

Determining the optimal capacity of facilities is regarded as one of the main strategic decisions in supply chain design problems. The importance of this strategic decision is more emphasized in biofuel supply chains compared to conventional supply chains. This is because the capital cost saving of larger facilities in biofuel supply chains that results from the economy of scale might be almost offset by the increased cost of transporting bulky biomass feedstocks for longer distances (Yue, You, & Snyder, 2014). The selection of capacity levels is modeled by binary variables, each of which corresponds to a specific capacity level and becomes equal to 1 if that level is chosen as the optimal level and 0 otherwise.

Depending on the type of biomass and targeting final products, there are various technological pathways for the conversion of biomass into biofuels.

The conversion technologies may vary in terms of operational and capital costs, conversion rate, energy consumption, and biomass feedstock requirements. Superstructure-based optimization is a powerful approach to construct a superstructure that embraces all feasible options for conversion units and all possible interconnections between them. The constructed superstructure is then translated into a mathematical programming problem that is solved to determine the optimal conversion pathway along with its operating conditions and design parameters (Yue & You, 2016). The reader is referred to Section 8.2 of Chapter 8 for more details on superstructure optimization.

3.2.4 Biofuel distribution network design

The current practice for biofuel delivery is to transport biofuels by tanker trucks from biorefineries to storage/distribution terminals where biofuels are blended in different proportions with their petroleum–derived fuel counterparts and then delivered to fueling stations by tanker trucks. The purpose of biofuel distribution networks is to cost-effectively deliver the produced biofuels in biorefineries to end users. Therefore, there is a need to systemically optimize and design the biofuel distribution network to determine the optimal number, location, and capacity of the terminals considering available candidate locations and transportation links.

For long-distance delivery of biofuels, however, tanker truck is not an economical option. In such cases, pipeline transportation that allows for the delivery of larger volumes of biofuel over longer distances can be used as a suitable transportation mode. As existing pipelines that are used for gasoline distribution are not appropriate for transporting biofuel due to its affinity for water and related corrosion problems (Ribeiro, 2013), dedicated pipeline networks must be designed, which can be introduced as a strategic decision into supply chain design models.

3.2.5 Integrated biofuel and petroleum supply chain design

Compared to conventional biofuels that are chemically and functionally distinct from petroleum fuels, hydrocarbon biofuels also called drop-in biofuels are chemically identical to petroleum gasoline, diesel, or jet fuel and are fully compatible with combustion engines and refinery infrastructures (Karatzos, Van Dyk, Mcmillan, & Saddler, 2017). This advantage provides a great opportunity for the biofuel industry to take benefit from the existing infrastructures, specifically pipelines and refinery units, resulting in a significant saving in the capital and operating costs. Three insertion points are recommended for liquid bio-intermediates (e.g., bio-slurry and bio-oil) produced from biomass to enter a petroleum refinery. In the first insertion point, bio-slurry produced from biomass is mixed with crude oil and then goes through crude distillation units (CDUs) and a series of upgrading units to produce gasoline, diesel, and jet fuel. In the second insertion point, bio-oil is directly fed to the upgrading units to produce the final products. In the third insertion point, biofuels are blended with their fossil fuels counterparts in petroleum refineries and then shipped to customers using the existing pipeline distribution networks (Tong, Gleeson, Rong, & You, 2014). The main strategic decisions in this regard include determining the optimal design of the hydrocarbon biofuel supply chains integrated with existing petroleum refineries as well as the identification of the optimal insertion point(s).

3.2.6 Biofuel market selection

The biofuels market can be segmented by type (ethanol and biodiesel), feedstock (forestry and agricultural residues, grains, starch crops, vegetable oils, waste oils and animal fats, dedicated energy crops, algae, and others), and geography. One of the important decisions that must be made before designing biofuel supply chains is which market segment to enter, taking into account factors affecting the biofuel market such as changes in government policies and incentives, uncertainty in costs and revenues associated with biofuel production, future competition of biofuels with conventional fossil fuels, possible adjustments to regional and national biofuel targets, and technology evolution.

3.2.7 International trade network design

International trade of biomass and biofuel seems to provide mutually beneficial opportunities for both importing and exporting countries. Importing biomass and biofuels from countries with lower materials, land, labor, and equipment costs in addition to domestic production could not only provide a more reliable supply of biofuels but also help biofuels become more price-competitive with fossil fuels and accelerate the sustainable development of the global biofuel industry.

Compared to domestic supply chains, global biofuel supply chains are more difficult to establish due to several reasons. While the cost of biomass production in some countries may be relatively low, long-distance biomass transportation between countries and even continents incurs additional logistic costs (Heinimö & Junginger, 2009). To increase the stability and energy density of biomass and save handling and transportation costs, bulky biomass undergoes pretreatment processes and is converted into a densified form before international transportation. Uncertainty in exchange rates, different local cultures, changes in the regulatory environment, infrastructure deficiencies in transportation and telecommunications in developing countries, political and economic instability, and geographical diversity are other factors that influence the performance of the global supply chains (Meixell & Gargeya, 2005). The additional management and planning challenges the global supply chains bring must be handled by a well-designed global supply chain, making international trade of biomass and biofuel feasible from economic and energy efficiency perspective.

3.2.8 Biofuel supply chain network redesign

Due to the large investments related to strategic decisions, these decisions are expected to last for a substantial duration of time. However, different types of changes during the lifetime of a facility may turn a desirable location in the present time into an undesirable one in the future. Therefore, taking into account the possibility of making adjustments in the structure of the supply chain and its related decisions seems to be essential in some cases (Melo, Nickel, & Saldanha-Da-Gama, 2009). Allowing for revising strategic decisions, including facilities number, location, capacity, and technology, as well as marketing and sourcing policies, at the beginning of planning time periods, leads to a class of design models, called dynamic supply chain network design problems (Klibi, Martel, & Guitouni, 2010). The supply chain network redesign problem that assumes the supply chain already exists deals with reconfiguration decisions such as facility relocation to new positions, opening of new facilities, and closing of existing facilities (Hammami & Frein, 2014).

The configuration of the biomass-to-biofuel supply chain is not static and can evolve over time because of issues such as technology development, new government incentives, and increased acceptance of biofuel products (Yue et al., 2014). Therefore, it is essential to consider the possibility of capacity expansion, and technology upgrading for existing biofuel supply chains. The main decisions addressed by biomass supply chain redesign models include determining whether the existing facilities must be kept open with the same capacity, expanded, or closed. Gonela, Zhang, Osmani, et al., 2015, for example, formulated the bioethanol supply chain redesign problem using binary variables to determine whether the existing biofuel plants should operate at the same capacity, expanded, or closed, as well as determining the location of new plants.

3.3 Tactical-level decisions

3.3.1 Harvesting and collection planning

Biomass harvesting and collection are challenging and costly operations in biofuel supply chains mainly due to the scattered geographical distribution and seasonality of biomass resources. Harvesting is necessary for some types of biomass that are not readily available such as dedicated energy crops (Malladi & Sowlati, 2018), and is done using three common procedures, namely, multi-pass, single-pass, and whole-crop (Sambra, Sørensen, & Kristensen, 2008). After being harvested, the biomass is collected and then stored or directly transported to biorefineries. There are three methods for collecting the harvested biomass, including baling, loafing, and dry chop. In the baling method, feedstocks are prepared in the form of rectangular or round bales of dry biomass to facilitate storage and transportation. In the loafing collection, dry biomass from windrow is compressed in the form of large stacks with a doom shape aiming at protecting biomass feedstocks from water. The dry chop method that is usually used for herbaceous plants with long stalks such as miscanthus chops biomass into small pieces and blows the chopped material into a forage wagon (Forsberg, 2000; Zandi Atashbar, Labadie, & Prins, 2018).

The above operations have substantial economic costs as well as environmental damages arising from the considerable energy demand of the machinery used to harvest and collect biomass. Therefore, the selection of an appropriate harvesting and collection method is of great importance to improve the economic and environmental sustainability of biofuel supply chains (San Miguel et al., 2015). This decision can be considered as a strategic decision that remains unchanged for several years, or as a tactical decision that is updated in each harvesting season in accordance with weather and climate conditions, yield characteristics, farmers feedback, and other received information. The other tactical decision that is taken in medium-term planning levels is determining the optimal quantity of biomass that must be harvested and collected from each supplier.

3.3.2 Biomass storage planning

One of the main tactical decisions associated with biomass storage is determining the seasonal or monthly inventory level of biomass in each storage center. The motivation for storing biomass is to balance its supply and demand because there is high regional and seasonal variability in biomass production while the demand for biofuels and biomass feedstocks must be satisfied continuously all year round. Storage decisions are addressed by multiple-period supply chain planning models where the inventory levels in consecutive periods are optimized such that the continuous supply of biomass is ensured.

The selection of biomass storage method is another decision that needs to be made in the medium-term planning level. Various methods have been proposed in the literature for storage of biomass, among which ambient storage covered with a plastic film is the cheapest option. However, the drawbacks include the lack of control over biomass moisture, significant material loss, and possible risks for human health. To avoid biomass quality degradation because of fermentation, infections, and material loss, more expensive methods such as closed warehouses with hot air injection and covered storage facilities with a pole-frame structure can be used (Rentizelas, Tolis, & Tatsiopoulos, 2009). It is clear that the tradeoff between material loss rate and storage costs must be considered in selecting the appropriate storage method.

3.3.3 Biomass and biofuel transportation planning

Due to their characteristics such as low energy density, spatial fragmentation, and high moisture content, biomass feedstocks are cost-prohibitive and inconsistent for long-distance transportation (De Meyer et al., 2014; Yue et al., 2014). Since expensive transportation and logistics costs are one of the key barriers slowing the development of the biofuel industry, transportation decisions are incorporated into most of the supply chain optimization models to ensure the optimal performance of transportation operations.

There are several transportation modes for moving material through the biofuel supply chains, including truck, barge, train, pipeline, or a combination of them (i.e., multimodal transportation), each of which can be chosen as a viable option depending on the type of biomass, volume of biomass transported, distance traveled, and other aspects. The transportation problems arising at the long-term and medium-term planning levels typically determine (1) which transportation mode (which combination of transportation modes) must be used, and (2) how much biomass and biofuel must be shipped by each transportation route and with which transportation route, while simultaneously pursuing the objectives of minimizing the direct costs of transportation as well as the environmental impacts of transportation.

3.3.4 Biorefinery process synthesis and design

Process synthesis and design, which is defined as the act of determining the optimal type and design of processing units as well as the optimal interconnection of the units within a biorefinery, allows the decision maker to design, synthesize, and retrofit biorefinery and its related conversion processes considering various criteria such as economics, sustainability, and efficiency (Andiappan, Ng, & Tan, 2017). There are two approaches to handle the problem of process synthesis and design. The first approach that is performed in a sequential manner fixes some components in the conversion processes and then adopts heuristic methods to generate possible changes that lead to the improvement of the performance of the biorefinery. The second approach that is based upon superstructure optimization postulates a superstructure that embraces all feasible options for the processing units and all possible interconnections between the units while eliminating unfeasible process flowsheets. The postulated superstructure is then translated into a mathematical programming problem. By solving the superstructure-based optimization model, all the process configurations are evaluated, and the optimal process flowsheet along with its operating conditions and design parameters is determined. The problem of process synthesis and design of biorefinery includes both strategic and tactical decisions, but its strategic decisions are more likely to need revision than those described in Section 3.2 (such as the location of biorefinery). The reason is that biomass conversion technologies are evolving and the established biorefineries must be retrofitted by adding new processes and technologies.

3.3.5 Supply chain master production planning

Biofuel supply chain master planning coordinates medium–term decisions related to biomass supply and storage, biomass transportation, biofuel production, and biofuel transportation and distribution with the purpose of meeting biofuel demands at minimum cost. The main decisions determined during the master production planning include the quantity of biomass cultivated, harvested, preprocessed, the quantity of biofuel produced, the quantity of biomass and biofuel stored, and the quantity of biomass and biofuel transported from supply areas to biorefineries and from biorefineries to demand zones, respectively. Traditionally, these decisions are made either independently or sequentially, leading to the poor overall performance of the supply chain. A supply chain master plan is generated centrally and adjusted periodically to synchronize the flow of biomass and biofuel along the chain. This enables the supply chain entities (i.e., biomass suppliers, storage units, and biorefineries) to reduce their inventory levels, while without coordination between the entities, larger buffers are required to mitigate the increased risks of stockouts and ensure a continuous flow of material. In other words, the master planning (Babazadeh et al., 2015) is responsible for generating an aggregated production and distribution plan for all entities of the supply chain based on the demand forecasts (Albrecht, Rohde, & Wagner, 2015; Fleischmann et al., 2005).

3.3.6 Biofuel pricing and biofuel demand forecasting

It is obvious that accurate forecasting models are required to predict future demand for biofuels. However, unpredictability of social aspects of biofuels and changing biofuel-related policies and regulations are among factors that lead to high volatility of biofuel demand, making biofuel demand forecasting more complex than demand forecasting for fossil fuels (Bairamzadeh, Saidi-Mehrabad, & Pishvaee, 2018).

While the price of petroleum products follows the price of crude oil as petroleum feedstock, the price of biofuels is coordinated to the price of their fossil fuel counterparts rather than biomass feedstocks. The lack of appropriate pricing policy of biomass and biofuels is one of the major barriers to the participation of farmers in changing their planting patterns and producing biomass feedstocks, resulting in the failure of the biofuel program in many parts of the world (Gracia, Velázquez-Martí, & Estornell, 2014; Pohit, Biswas, Kumar, & Goswami, 2010). Therefore, there is a need for supply chain planning models to develop dynamic pricing and incentive mechanisms that ensure sufficient motivation for noncooperative farmers to cooperate and for manufacturers to invest in biorefineries (Bai, Ouyang, & Pang, 2012).

3.4 Operational-level decisions

3.4.1 Scheduling of supply chain operations

One important category of the operational-level decisions is the detailed planning and scheduling of all operations involved in biofuel supply chains, including biomass harvest and collection, biomass preprocessing and storage, biomass transportation, biomass-to-biofuel conversion, biofuel transportation and distribution, and biofuel sale.

Biomass harvesting and collection consist of several sequential tasks (e.g., cutting, raking, baling, and loading) that must be scheduled on different types of machines in several geographically distributed biomass fields. Due to the seasonal availability of biomass and its short harvest interval, harvest season of many types of biomass may overlap. Therefore, it is necessary to carefully allocate harvest equipment to the biomass fields and optimize the routing of equipment between them such that the harvest of all the fields is completed within their allowable time window. In the biomass supply and logistics literature, the biomass harvesting and collection problem has been cast as well-known combinatorial optimization problems such as traveling salesman problem (TSP), machine scheduling problem, and vehicle routing problem (VRP) with time windows aiming at determining the optimal sequence in which the fields are visited. Basnet, Foulds, and Wilson (2006) presented a harvest scheduling model for a contracting company traveling from field to field to harvest crops. The proposed model formulated as the TSP determined the activity of each machine and its crew in different time periods while simultaneously minimizing the duration of the entire harvesting process. Bochtis et al. (2013) addressed the problem of finding an optimal schedule for collection operations (i.e., balling and loading) in a number of fields. They assumed that two machinery systems work sequentially for cotton residues collection, including baling (i.e., tractor-baler) and loading (i.e., fork-lift or transport truck) systems, and formulated this problem as a flow shop scheduling problem with the sequence-dependent set up times considering the objective function of minimizing the total completion time of all tasks in all fields. Gracia et al. (2014) addressed the residual agricultural biomass harvesting and collection problem as a VRP with the objective of minimizing the costs of routes calculated for a fleet of multiple agricultural vehicles, including trucks, chippers, tractors, and tipper trailers.

Production planning and scheduling concerns the assignment of resources such as biomass feedstocks, raw materials, utilities, and equipment units to biofuel production tasks as well as the sequencing and timing of the tasks on processing units with the view of optimizing one or more objectives like the minimization of the total cost, lateness/earliness, and makespan (Maravelias & Sung, 2009). Another operational-level decision that is often accompanied by production scheduling is related to inventory decisions that control how much biomass is (1) added to storage centers, (2) lost during storage, and (3) utilized for biofuel production, as well as how much biofuel is (1) produced and stored in biorefineries, and (2) delivered to the downstream.

3.4.2 Vehicle routing and scheduling

Operational-level transportation planning copes with the movement of material between different locations throughout the biofuel supply chain over a short-term horizon, typically a day or a week. The seasonal and scattered availability of biomass together with its low energy density make the transportation of biomass more complex and costly than the transportation of high energy density biofuels that can be handled by companies having long-term experience with fossil fuel transportation. Operational decisions corresponding to the transportation include vehicle routing and fleet scheduling to perform biomass and biofuel pickup and delivery activities. Han and Murphy (2012), as an example, addressed the problem of truck scheduling for shipping forest woody biomass in western Oregon. They presented a mathematical model with two objectives of minimizing the working time for a whole day and the total transportation cost, considering constraints related to transportation routes, working time or labor, and predetermined order requirements. Torjai and Kruzslicz (2016) introduced the biomass truck scheduling problem for the herbaceous-based biofuel supply chain. The authors considered the truck scheduling problem as a parallel machine scheduling where parallel machines (i.e., identical trucks) deliver biomass from storage centers to a central biorefinery. They proposed mixed-integer programming mathematical models to deal with the scheduling problem while minimizing the number of trucks required for biomass transportation and their total idle times.

3.5 Systematic classification of the literature

In order to classify and review the existing literature of the biomass-to-biofuel supply chain design and planning models, the relevant studies are categorized based on the classification criteria shown in Table 3.1. Table 3.2 provides the classification of the related literature based on the decision-making levels, including strategic, tactical, and operational, as well as

Table 3.1 Classification criteria and abbreviations

Strategic decisions

selection of biomass types	SB	Selection of cultivation sites	SC
Biorefinery location selection	FL	Biorefinery capacity selection	FC
Biorefinery technology selection	TS	Capacity expansion or reduction/Technology upgrading	C/T
Biofuel distribution network design/ Integrated biofuel and petroleum supply chain design	D/I	Market selection/International trade network design	M/I

Tactical decisions

Biofuel production planning/ supply chain master production planning	P/S	Biorefinery process synthesis and design	PS
Biomass and biofuel transportation planning	TP	Inventory planning	IP
Harvest and collection planning	HC	Biofuel pricing/Demand forecasting	B/D

Operational decisions

Scheduling of harvest and collection	HC	Scheduling and routing of vehicles for biomass transportation	VB
Scheduling of conversion tasks and processing units	CP	Scheduling and routing of vehicles for biofuel transportation	VF

Table 3.2 Classification of recent publications on biofuel supply chain design and planning

Reference	Strategic decisions								Tactical decisions						Operational decisions			
	SB	SC	FL	FC	TS	C/T	D/I	M/I	B/S	TP	IP	HC	PS	B/D	HC	RV	CP	VF
Graham et al. (200C)		✓	✓	✓					✓	✓		✓			✓			
Tatsiopoulos and Tclis (2003)																✓		
Basnet et al. (2006)																		
Sokhansanj et al. (2006)															✓✓	✓		
Ekşioğlu et al. (2009)			✓	✓					✓	✓	✓							
Rentizelas et al. (2009)			✓	✓					✓	✓	✓							
Guillén-Gosálbez ard Grossmann (2009)			✓	✓					✓	✓	✓		✓	✓				
Leduc et al. (2009)			✓	✓		✓						✓						
Pohit et al. (2010)		✓	✓	✓	✓	✓												
Baliban et al. (2013)		✓	✓	✓	✓									✓				
Dal-Mas et al. (2011)			✓	✓✓	✓				✓✓	✓✓								
Kim et al. (2011)			✓	✓														
Healey et al. (2011)			✓	✓✓		✓			✓✓	✓✓								
Akgul et al. (2012)		✓	✓✓	✓✓	✓				✓✓	✓✓	✓							
You and Wang (2011)					✓													
An and Searcy (2012)									✓✓	✓✓								
Han and Murphy (2012)								✓						✓		✓		
Bai et al. (2012)		✓	✓✓	✓✓	✓✓				✓✓✓	✓✓✓	✓✓							
Chen and Fan (2012)		✓	✓✓	✓✓					✓✓✓	✓✓✓								
Gebreslassie et al. (2C12)			✓	✓	✓													
Kostin et al. (2012)			✓	✓														

(Continued)

Table 3.2 Classification of recent publications on biofuel supply chain design and planning (*Cont.*)

Reference	Strategic decisions								Tactical decisions						Operational decisions			
	SB	SC	FL	FC	TS	C/T	D/I	M/I	B/S	TP	IP	HC	PS	B/D	HC	RV	CP	VF
Marvin et al. (2012)			✓	✓					✓	✓								
Walther et al. (2012)			✓	✓	✓				✓	✓								
Awudu and Zhang (2013)						✓			✓	✓	✓							
Giarola et al. (2013)	✓	✓	✓	✓	✓				✓	✓		✓						
Sharma et al. (2013a)				✓						✓								
Osmani and Zhang (2013)		✓	✓		✓				✓	✓	✓							
Bochtis et al. (2013)					✓										✓			
Murillo–Alvarado et al. (2013)		✓					✓						✓		✓			
Orfanou et al. (2013)			✓	✓	✓		✓								✓			
Gracia et al. (2014)			✓	✓	✓				✓	✓								
Tong et al. (2014a)			✓	✓					✓	✓	✓							
Tong et al. (2014b)			✓	✓					✓	✓	✓							
Li and Hu (2014)									✓	✓	✓						✓	
Marufuzzaman et al. (2014)	✓									✓	✓							
Azadeh et al. (2014)			✓	✓	✓				✓	✓								
Tong et al. (2014c)		✓	✓	✓					✓	✓								
Balaman and Selim (2014)			✓	✓	✓				✓	✓								

Reference	Strategic decisions								Tactical decisions						Operational decisions			
	SB	SC	FL	FC	TS	C/T	D/I	M/I	B/S	TP	IP	HC	PS	B/D	HC	RV	CP	VF
Kelloway and Daoutidis (2014)		✓	✓		✓								✓					
Osmani and Zhang (2014)			✓	✓														
Gonela et al. (2015a)		✓	✓	✓					✓	✓	✓							
Gonela et al. (2015b)		✓	✓	✓					✓	✓	✓							
Babazadeh et al. (2015)		✓	✓		✓				✓	✓	✓							
Bairamzadeh et al. (2016)			✓	✓	✓				✓	✓	✓							
Azadeh and Arani (2016)			✓	✓	✓				✓	✓	✓							
Santibañez-Aguilar et al. (2016)						✓			✓	✓	✓							
Shabani and Sowlati (2016)						✓												
Mohseni and Pishvaee (2016)			✓	✓	✓													
Mohseni et al. (2016)		✓	✓	✓	✓				✓	✓	✓							
Caffrey et al. (2015)		✓	✓									✓				✓		
Torjai and Kruzslicz (2016)																✓		
Najmi et al. (2016)			✓	✓	✓				✓		✓							
How et al. (2016)			✓	✓	✓				✓		✓							
Babazadeh et al. (2017)		✓	✓	✓					✓	✓	✓			✓				
Bairamzadeh et al. (2018)			✓	✓					✓	✓	✓							
Huang and Hu (2018)		✓	✓	✓							✓							
Mirhashemi et al. (2018)		✓	✓	✓	✓						✓							
Nodooshan et al. (2018)																		

(Continued)

Table 3.2 Classification of recent publications on biofuel supply chain design and planning (Cont.)

Reference	Strategic decisions								Tactical decisions						Operational decisions			
	SB	SC	FL	FC	TS	C/T	D/I	M/I	B/S	TP	IP	HC	PS	B/D	HC	RV	CP	VF
Abasian et al. (2019)	✓																	
Razm et al. (2019)			✓	✓	✓				✓	✓								
Mohseni and Pishvaee (2020)			✓	✓					✓	✓		✓						
Ahranjani et al. (2020)			✓	✓	✓				✓	✓	✓							

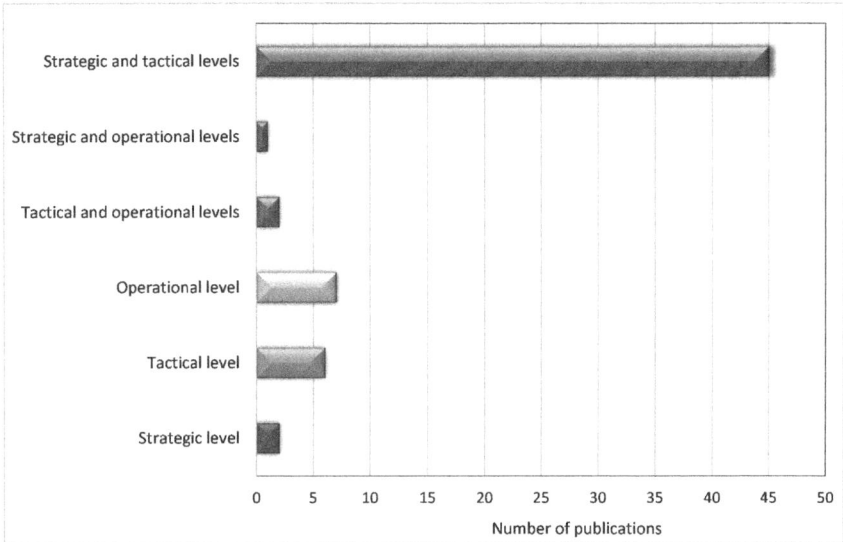

Figure 3.2 *Share of strategic, tactical, and operational decisions in the biofuel supply chain studies.*

corresponding decision variables addressed at each decision level. The main conclusions and research gaps identified through the literature review are presented as follows:

- As depicted in Fig. 3.2, most of the researches have focused on strategic and tactical decisions, while less emphasis has been placed on investigating operational decisions and combined decisions of operational with strategic or tactical levels.
- According to Fig. 3.3 that represents the distribution of various strategic decisions addressed in biomass supply chain literature, determination of location, capacity, and technology of biorefineries have obtained much attention compared to other strategic decisions in the literature. More specifically, a few papers have addressed capacity expansion/reduction, technology upgrading, and particularly supply chain network redesign problems. On the other hand, most of the optimization models considering biomass sourcing decisions have addressed the selection of cultivation sites rather than biomass sourcing strategy and selection of biomass types decisions. Moreover, few papers have investigated biofuel distribution network design, integrated biofuel and petroleum supply chain design, biofuel market selection, and international trade of biofuel in the literature.
- Based on Fig. 3.4, tactical decisions that have been investigated in a considerable amount of the related studies include biomass allocation,

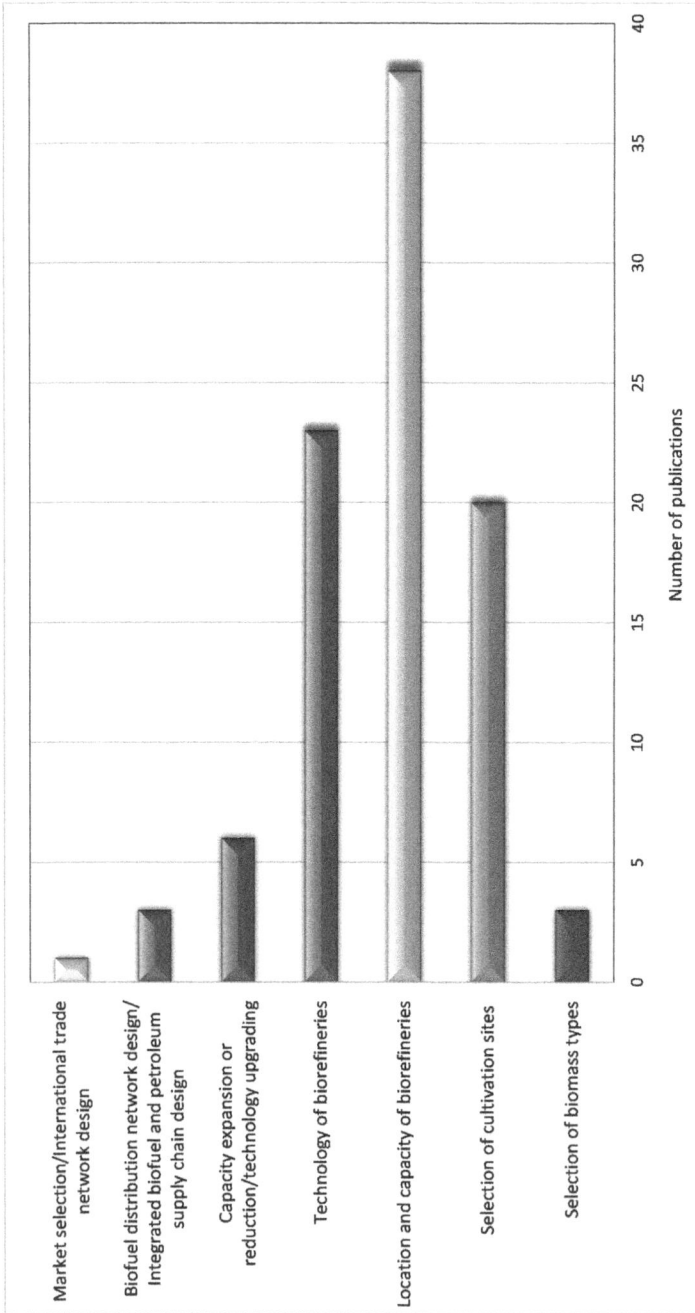

Figure 3.3 Share of different strategic decisions considered in the biofuel supply chain studies.

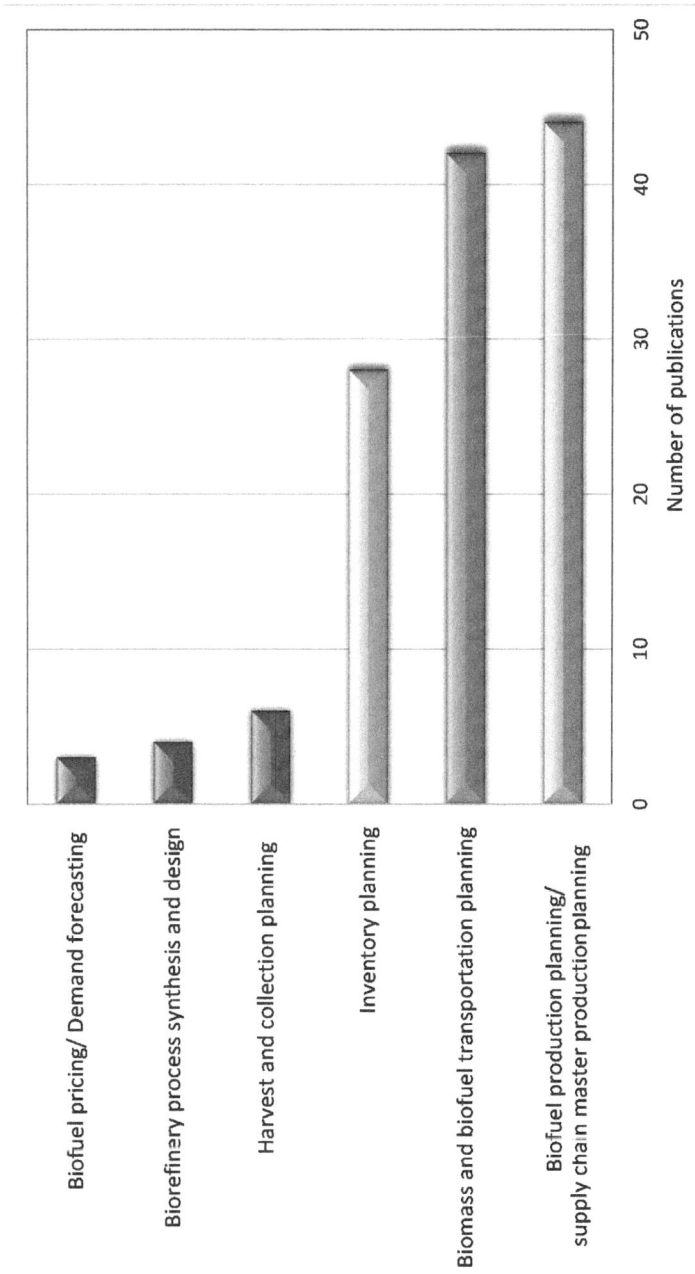

Figure 3.4 *Share of different tactical decisions considered in the biofuel supply chain studies.*

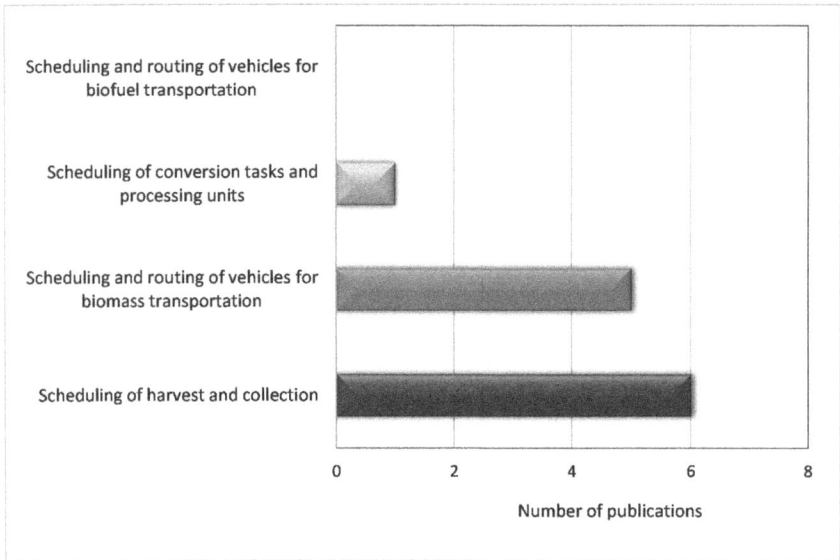

Figure 3.5 *Share of different operational decisions considered in the biofuel supply chain studies.*

biofuel production, distribution planning of biomass/biofuel, and inventory planning. On the other hand, less attention has been paid to decisions about the selection of storage methods, biofuel pricing and demand forecasting, and biomass supply contracts.

- As mentioned before and based on Fig. 3.5, a small fraction of the previous studies has been devoted to the optimization of operational decisions and a combination of operational decisions with tactical or strategic decisions. In this regard, most of these models have investigated the operational decisions in the biomass supply stage, including scheduling and routing of harvest and collection equipment, allocation of labors and equipment, arrangement of working shifts, and scheduling and routing of vehicles for biomass transportation. Accordingly, studies investigating scheduling of conversion tasks and processing units, allocation of raw materials, labor, and equipment units, scheduling and routing of vehicles for biofuel transportation, and biofuel and byproducts demand fulfillment are rare.

3.6 Conclusions

This chapter provides a framework for biomass-to-biofuel supply chain planning decisions, which categorizes the planning tasks of the chain in two dimensions of planning horizons and supply chain stages. Planning horizons incorporate strategic, tactical, and operational decision levels, while supply chain processes consist of biomass supply and preprocessing, biofuel production, biofuel blending and distribution, and biofuel sales stages. A detailed review of the biomass supply chain design and planning models reveals that the strategic and tactical decisions have obtained much attention in the literature, whilst less emphasis has been devoted to operational decisions. In this regard, among strategic and decisions, several issues such as biomass sourcing strategy and selection of biomass types, supply chain network redesign, blending strategy selection, market selection, international trade of biofuel, biofuel pricing, and demand forecasting have been investigated in few papers. On the other hand, the biofuel supply chain literature has rarely studied some operational decisions, including scheduling of conversion tasks and processing units, allocation of raw materials, labor, and equipment units, scheduling and routing of vehicles for biofuel transportation, and biofuel and byproducts demand fulfillment, which can be a direction for future researches.

References

Abasian, F., Rönnqvist, M., & Ouhimmou, M. (2019). Forest bioenergy network design under market uncertainty. *Energy*, *188*, 116038.

Ahranjani, P. M., Ghaderi, S. F., Azadeh, A., & Babazadeh, R. (2020). Robust design of a sustainable and resilient bioethanol supply chain under operational and disruption risks. *Clean Technologies and Environmental Policy*, *22*, 119–151.

Akgul, O., Shah, N., & Papageorgiou, L. G. (2012). An optimisation framework for a hybrid first/second generation bioethanol supply chain. *Computers & Chemical Engineering*, *42*, 101–114.

Albrecht, M., Rohde, J., & Wagner, M. (2015). *Master planning. Supply chain management and advanced planning*. Berlin, Heidelberg: Springer, 155–175.

An, H., & Searcy, S. W. (2012). Economic and energy evaluation of a logistics system based on biomass modules. *Biomass and Bioenergy*, *46*, 190–202.

Anderson, E., Arundale, R., Maughan, M., Oladeinde, A., Wycislo, A., & Voigt, T. (2011). Growth and agronomy of Miscanthus × giganteus for biomass production. *Biofuels*, *2*, 71–87.

Andiappan, V., Ng, D. K., & Tan, R. R. (2017). Design operability and retrofit analysis (DORA) framework for energy systems. *Energy*, *134*, 1038–1052.

Awudu, I., & Zhang, J. (2013). Stochastic production planning for a biofuel supply chain under demand and price uncertainties. *Applied Energy*, *103*, 189–196.

Azadeh, A., & Arani, H. V. (2016). Biodiesel supply chain optimization via a hybrid system dynamics-mathematical programming approach. *Renewable Energy*, *93*, 383–403.

Azadeh, A., Arani, H. V., & Dashti, H. (2014). A stochastic programming approach towards optimization of biofuel supply chain. *Energy*, *76*, 513–525.

Babazadeh, R., Razmi, J., Pishvaee, M. S., & Rabbani, M. (2015). A non-radial DEA model for location optimization of *Jatropha curcas* L. cultivation. *Industrial Crops and Products*, *69*, 197–203.

Babazadeh, R., Razmi, J., Rabbani, M., & Pishvaee, M. S. (2017). An integrated data envelopment analysis–mathematical programming approach to strategic biodiesel supply chain network design problem. *Journal of Cleaner Production*, *147*, 694–707.

Bai, Y., Ouyang, Y., & Pang, J. -S. (2012). Biofuel supply chain design under competitive agricultural land use and feedstock market equilibrium. *Energy Economics*, *34*, 1623–1633.

Bairamzadeh, S., Pishvaee, M. S., & Saidi-Mehrabad, M. (2016). Multiobjective robust possibilistic programming approach to sustainable bioethanol supply chain design under multiple uncertainties. *Industrial & Engineering Chemistry Research*, *55*, 237–256.

Bairamzadeh, S., Saidi-Mehrabad, M., & Pishvaee, M. S. (2018). Modelling different types of uncertainty in biofuel supply network design and planning: A robust optimization approach. *Renewable Energy*, *116*, 500–517.

Balaman, Ş. Y., & Selim, H. (2014). A fuzzy multiobjective linear programming model for design and management of anaerobic digestion based bioenergy supply chains. *Energy*, *74*, 928–940.

Baliban, R. C., Elia, J. A., & Floudas, C. A. (2013). Biomass to liquid transportation fuels (BTL) systems: Process synthesis and global optimization framework. *Energy & Environmental Science*, *6*, 267–287.

Basnet, C. B., Foulds, L. R., & Wilson, J. M. (2006). Scheduling contractors' farm-to-farm crop harvesting operations. *International Transactions in Operational Research*, *13*, 1–15.

Bochtis, D., Dogoulis, P., Busato, P., Sørensen, C., Berruto, R., & Gemtos, T. (2013). A flow-shop problem formulation of biomass handling operations scheduling. *Computers and Electronics in Agriculture*, *91*, 49–56.

Caffrey, K., Chinn, M., Veal, M., & Kay, M. (2015). Biomass supply chain management in North Carolina (part 2): Biomass feedstock logistical optimization. *AIMS Energy*, *4*, 280–299.

Chen, C. -W., & Fan, Y. (2012). Bioethanol supply chain system planning under supply and demand uncertainties. *Transportation Research Part E: Logistics and Transportation Review*, *48*, 150–164.

Dal-Mas, M., Giarola, S., Zamboni, A., & Bezzo, F. (2011). Strategic design and investment capacity planning of the ethanol supply chain under price uncertainty. *Biomass and Bioenergy*, *35*, 2059–2071.

De Meyer, A., Cattrysse, D., Rasinmäki, J., & Van Orshoven, J. (2014). Methods to optimise the design and management of biomass-for-bioenergy supply chains: A review. *Renewable and Sustainable Energy Reviews*, *31*, 657–670.

Ekşioğlu, S. D., Acharya, A., Leightley, L. E., & Arora, S. (2009). Analyzing the design and management of biomass-to-biorefinery supply chain. *Computers & Industrial Engineering*, *57*, 1342–1352.

Fleischmann, B., Meyr, H., & Wagner, M. (2005). *Advanced planning. Supply chain management and advanced planning*. Berlin, Heidelberg: Springer, 81–106.

Forsberg, G. (2000). Biomass energy transport: Analysis of bioenergy transport chains using life cycle inventory method. *Biomass and Bioenergy*, *19*, 17–30.

Gebreslassie, B. H., Yao, Y., & You, F. (2012). Design under uncertainty of hydrocarbon biorefinery supply chains: Multiobjective stochastic programming models, decomposition algorithm, and a comparison between CVaR and downside risk. *AIChE Journal*, *58*, 2155–2179.

Giarola, S., Bezzo, F., & Shah, N. (2013). A risk management approach to the economic and environmental strategic design of ethanol supply chains. *Biomass and Bioenergy*, *58*, 31–51.

Gonela, V., Zhang, J., & Osmani, A. (2015a). Stochastic optimization of sustainable industrial symbiosis based hybrid generation bioethanol supply chains. *Computers & Industrial Engineering*, *87*, 40–65.

Gonela, V., Zhang, J., Osmani, A., & Onyeaghala, R. (2015b). Stochastic optimization of sustainable hybrid generation bioethanol supply chains. *Transportation Research Part E: Logistics and Transportation Review*, *77*, 1–28.

Gracia, C., Velázquez-Martí, B., & Estornell, J. (2014). An application of the vehicle routing problem to biomass transportation. *Biosystems Engineering*, *124*, 40–52.

Graham, R. L., English, B. C., & Noon, C. E. (2000). A geographic information system-based modeling system for evaluating the cost of delivered energy crop feedstock. *Biomass and Bioenergy*, *18*, 309–329.

Guillén-Gosálbez, G., & Grossmann, I. E. (2009). Optimal design and planning of sustainable chemical supply chains under uncertainty. *AIChE Journal, 55*, 99–121.

Hammami, R., & Frein, Y. (2014). Redesign of global supply chains with integration of transfer pricing: Mathematical modeling and managerial insights. *International Journal of Production Economics, 158*, 267–277.

Han, S. -K., & Murphy, G. E. (2012). Solving a woody biomass truck scheduling problem for a transport company in Western Oregon, USA. *Biomass and Bioenergy, 44*, 47–55.

Healey, N. C., Irmak, A., Hubbard, K. G., & Lenters, J. D. (2011). Environmental variables controlling site suitability for corn-based ethanol production in Nebraska. *Biomass and Bioenergy, 35*, 2852–2860.

Heinimö, J., & Junginger, M. (2009). Production and trading of biomass for energy – An overview of the global status. *Biomass and Bioenergy, 33*, 1310–1320.

How, B. S., Tan, K. Y., & Lam, H. L. (2016). Transportation decision tool for optimisation of integrated biomass flow with vehicle capacity constraints. *Journal of Cleaner Production, 136*, 197–223.

Huang, S., & Hu, G. (2018). Biomass supply contract pricing and environmental policy analysis: A simulation approach. *Energy, 145*, 557–566.

Karatzos, S., Van Dyk, J. S., Mcmillan, J. D., & Saddler, J. (2017). Drop-in biofuel production via conventional (lipid/fatty acid) and advanced (biomass) routes. Part I. *Biofuels, Bioproducts and Biorefining, 11*, 344–362.

Kelloway, A., & Daoutidis, P. (2014). Process synthesis of biorefineries: Optimization of biomass conversion to fuels and chemicals. *Industrial & Engineering Chemistry Research, 53*, 5261–5273.

Kim, J., Realff, M. J., Lee, J. H., Whittaker, C., & Furtner, L. (2011). Design of biomass processing network for biofuel production using an MILP model. *Biomass and Bioenergy, 35*, 853–871.

Klibi, W., Martel, A., & Guitouni, A. (2010). The design of robust value-creating supply chain networks: A critical review. *European Journal of Operational Research, 203*, 283–293.

Kostin, A., Guillén-Gosálbez, G., Mele, F., Bagajewicz, M., & Jiménez, L. (2012). Design and planning of infrastructures for bioethanol and sugar production under demand uncertainty. *Chemical Engineering Research and Design, 90*, 359–376.

Langholtz, M., Webb, E., Preston, B. L., Turhollow, A., Breuer, N., Eaton, L., King, A. W., Sokhansanj, S., Nair, S. S., & Downing, M. (2014). Climate risk management for the US cellulosic biofuels supply chain. *Climate Risk Management, 3*, 96–115.

Leduc, S., Natarajan, K., Dotzauer, E., Mccallum, I., & Obersteiner, M. (2009). Optimizing biodiesel production in India. *Applied Energy, 86*, S125–S131.

Li, Q., & Hu, G. (2014). Supply chain design under uncertainty for advanced biofuel production based on bio-oil gasification. *Energy, 74*, 576–584.

Malladi, K. T., & Sowlati, T. (2018). Biomass logistics: A review of important features, optimization modeling and the new trends. *Renewable and Sustainable Energy Reviews, 94*, 587–599.

Maravelias, C. T., & Sung, C. (2009). Integration of production planning and scheduling: Overview, challenges and opportunities. *Computers & Chemical Engineering, 33*, 1919–1930.

Marufuzzaman, M., Eksioglu, S. D., & Huang, Y. E. (2014). Two-stage stochastic programming supply chain model for biodiesel production via wastewater treatment. *Computers & Operations Research, 49*, 1–17.

Marvin, W. A., Schmidt, L. D., Benjaafar, S., Tiffany, D. G., & Daoutidis, P. (2012). Economic optimization of a lignocellulosic biomass-to-ethanol supply chain. *Chemical Engineering Science, 67*, 68–79.

Meixell, M. J., & Gargeya, V. B. (2005). Global supply chain design: A literature review and critique. *Transportation Research Part E: Logistics and Transportation Review, 41*, 531–550.

Melo, M. T., Nickel, S., & Saldanha-Da-Gama, F. (2009). Facility location and supply chain management – A review. *European Journal of Operational Research, 196*, 401–412.

Mirhashemi, M. S., Mohseni, S., Hasanzadeh, M., & Pishvaee, M. S. (2018). Moringa oleifera biomass-to-biodiesel supply chain design: An opportunity to combat desertification in Iran. *Journal of Cleaner Production, 203*, 313–327.

Mohseni, S., & Pishvaee, M. S. (2016). A robust programming approach towards design and optimization of microalgae-based biofuel supply chain. *Computers & Industrial Engineering, 100*, 58–71.

Mohseni, S., & Pishvaee, M. S. (2020). Data-driven robust optimization for wastewater sludge-to-biodiesel supply chain design. *Computers & Industrial Engineering, 139*, 105944.

Mohseni, S., Pishvaee, M. S., & Sahebi, H. (2016). Robust design and planning of microalgae biomass-to-biodiesel supply chain: A case study in Iran. *Energy, 111*, 736–755.

Murillo-Alvarado, P. E., Ponce-Ortega, J. M., Serna-Gonzalez, M., Castro-Montoya, A. J., & El-Halwagi, M. M. (2013). Optimization of pathways for biorefineries involving the selection of feedstocks, products, and processing steps. *Industrial & Engineering Chemistry Research, 52*, 5177–5190.

Myllyviita, T., Holma, A., Antikainen, R., Lähtinen, K., & Leskinen, P. (2012). Assessing environmental impacts of biomass production chains–application of life cycle assessment (LCA) and multi-criteria decision analysis (MCDA). *Journal of Cleaner Production, 29*, 238–245.

Najmi, A., Shakouri, H., & Nazari, S. (2016). An integrated supply chain: A large scale complementarity model for the biofuel markets. *Biomass and Bioenergy, 86*, 88–104.

Nanda, S., Dalai, A. K., & Kozinski, J. A. (2016). Supercritical water gasification of timothy grass as an energy crop in the presence of alkali carbonate and hydroxide catalysts. *Biomass and Bioenergy, 95*, 378–387.

Nodooshan, K. G., Moraga, R. J., Chen, S. -J. G., Nguyen, C., Wang, Z., & Mohseni, S. (2018). Environmental and economic optimization of algal biofuel supply chain with multiple technological pathways. *Industrial & Engineering Chemistry Research, 57*, 6910–6925.

Orfanou, A., Busato, P., Bochtis, D., Edwards, G., Pavlou, D., Sørensen, C. G., & Berruto, R. (2013). Scheduling for machinery fleets in biomass multiple-field operations. *Computers and Electronics in Agriculture, 94*, 12–19.

Osmani, A., & Zhang, J. (2013). Stochastic optimization of a multi-feedstock lignocellulosic-based bioethanol supply chain under multiple uncertainties. *Energy, 59*, 157–172.

Osmani, A., & Zhang, J. (2014). Economic and environmental optimization of a large scale sustainable dual feedstock lignocellulosic-based bioethanol supply chain in a stochastic environment. *Applied Energy, 114*, 572–587.

Pishvaee, M. S., Farahani, R. Z., & Dullaert, W. (2010). A memetic algorithm for bi-objective integrated forward/reverse logistics network design. *Computers & Operations Research, 37*, 1100–1112.

Pohit, S., Biswas, P. K., Kumar, R., & Goswami, A. (2010). Pricing model for biodiesel feedstock: A case study of Chhattisgarh in India. *Energy Policy, 38*, 7487–7496.

Razm, S., Nickel, S., Saidi-Mehrabad, M., & Sahebi, H. (2019). A global bioenergy supply network redesign through integrating transfer pricing under uncertain condition. *Journal of Cleaner Production, 208*, 1081–1095.

Rentizelas, A. A., Tolis, A. J., & Tatsiopoulos, I. P. (2009). Logistics issues of biomass: The storage problem and the multi-biomass supply chain. *Renewable and Sustainable Energy Reviews, 13*, 887–894.

Ribeiro, B. E. (2013). Beyond commonplace biofuels: Social aspects of ethanol. *Energy Policy, 57*, 355–362.

Sambra, A., Sørensen, C., & Kristensen, E. F. (2008). *Optimized harvest and logistics for biomass supply chain.* Valencia, Spain (DVD): Proceedings of European Biomass Conference and Exhibition. ISBN-10:8889407581, ISBN-13,2008.978-8889407585.

San Miguel, G., Corona, B., Ruiz, D., Landholm, D., Laina, R., Tolosana, E., Sixto, H., & Cañellas, I. (2015). Environmental, energy and economic analysis of a biomass supply chain based on a poplar short rotation coppice in Spain. *Journal of Cleaner production*, *94*, 93–101.

Santibañez-Aguilar, J. E., Morales-Rodriguez, R., González-Campos, J. B., & Ponce-Ortega, J. M. (2016). Stochastic design of biorefinery supply chains considering economic and environmental objectives. *Journal of Cleaner Production*, *136*, 224–245.

Shabani, N., & Sowlati, T. (2016). A hybrid multi-stage stochastic programming-robust optimization model for maximizing the supply chain of a forest-based biomass power plant considering uncertainties. *Journal of Cleaner Production*, *112*, 3285–3293.

Sharma, B., Ingalls, R. G., Jones, C. L., Huhnke, R. L., & Khanchi, A. (2013a). Scenario optimization modeling approach for design and management of biomass-to-biorefinery supply chain system. *Bioresource Technology*, *150*, 163–171.

Sharma, B., Ingalls, R. G., Jones, C. L., & Khanchi, A. (2013b). Biomass supply chain design and analysis: Basis, overview, modeling, challenges, and future. *Renewable and Sustainable Energy Reviews*, *24*, 608–627.

Sokhansanj, S., Kumar, A., & Turhollow, A. F. (2006). Development and implementation of integrated biomass supply analysis and logistics model (IBSAL). *Biomass and Bioenergy*, *30*, 838–847.

Tatsiopoulos, I., & Tolis, A. (2003). Economic aspects of the cotton-stalk biomass logistics and comparison of supply chain methods. *Biomass and Bioenergy*, *24*, 199–214.

Tong, K., Gleeson, M. J., Rong, G., & You, F. (2014a). Optimal design of advanced drop-in hydrocarbon biofuel supply chain integrating with existing petroleum refineries under uncertainty. *Biomass and Bioenergy*, *60*, 108–120.

Tong, K., Gong, J., Yue, D., & You, F. (2014b). Stochastic programming approach to optimal design and operations of integrated hydrocarbon biofuel and petroleum supply chains. *ACS Sustainable Chemistry & Engineering*, *2*, 49–61.

Tong, K., You, F., & Rong, G. (2014c). Robust design and operations of hydrocarbon biofuel supply chain integrating with existing petroleum refineries considering unit cost objective. *Computers & Chemical Engineering*, *68*, 128–139.

Torjai, L., & Kruzslicz, F. (2016). Mixed integer programming formulations for the biomass truck scheduling problem. *Central European Journal of Operations Research*, *24*, 731–745.

Walther, G., Schatka, A., & Spengler, T. S. (2012). Design of regional production networks for second generation synthetic bio-fuel – A case study in Northern Germany. *European Journal of Operational Research*, *218*, 280–292.

You, F., & Wang, B. (2011). Life cycle optimization of biomass-to-liquid supply chains with distributed–centralized processing networks. *Industrial & Engineering Chemistry Research*, *50*, 10102–10127.

Yue, D., & You, F. (2016). *Biomass and biofuel supply chain modeling and optimization. Biomass supply chains for bioenergy and biorefining.* Woodhead Publishing, 149–166.

Yue, D., You, F., & Snyder, S. W. (2014). Biomass-to-bioenergy and biofuel supply chain optimization: Overview, key issues and challenges. *Computers & Chemical Engineering*, *66*, 36–56.

Zandi Atashbar, N., Labadie, N., & Prins, C. (2018). Modelling and optimisation of biomass supply chains: A review. *International Journal of Production Research*, *56*, 3482–3506.

CHAPTER 4

Uncertainties in biofuel supply chain

4.1 Biofuel supply chain risk management framework

To enable the large scale and cost-competitive production of biofuels, it is of great importance to develop a supply chain network design model that systematically designs and optimizes the entire biomass-to-biofuel supply chain from biomass supply sites to consumption markets. Uncertainty is considered as one of the most crucial factors influencing the performance of the supply chain and, if ignored, can lead to suboptimal performance or even infeasible supply chain configurations (Shabani and Sowlati, 2016). Biomass supply chains are subject to various sources of uncertainty. In the upstream of the supply chain, biomass production is qualitatively and quantitatively affected by weather and climate conditions, making biomass supply vary from one season to another, from one year to another, and from one location to another (Poudel, Marufuzzaman, & Bian, 2016). Other upstream uncertainties are related to biomass storage, biomass price, and biomass transportations. In biorefineries, there are several uncertainties such as capital and operating cost uncertainty, technology uncertainty, and production rate uncertainty, mainly arising from the fact that the biofuel industry is still in its infancy and large-scale commercial biorefineries are not in place yet (Yue, You, & Snyder, 2014). Further down the supply chain, the biofuel market is surrounded by demand and price uncertainties and is sensitive to economic and financial fluctuations. In addition to the operational risks mentioned above, disruption risks originating from man-made or natural disasters (e.g., terrorist attacks, earthquakes, droughts, and floods) threaten the structure and operation of the supply chain. The 2012 drought experienced throughout the United States, for example, acknowledges the vulnerability of biofuel production to such extreme events (Langholtz et al., 2014). In order to ensure that the supply chain model is effective and practical for real-world applications, it must provide an adequate safeguard against uncertainty (Mohseni, Pishvaee, & Sahebi, 2016). To achieve this goal, there is a need to develop a comprehensive risk management framework that can be used for supply chain optimization models under uncertainty. During

Biomass to Biofuel Supply Chain Design and Planning under Uncertainty
http://dx.doi.org/10.1016/B978-0-12-820640-9.00004-0
65

the last two decades, a number of conceptual frameworks have been proposed in the supply chain risk management literature to help business and industrial sectors handle supply chain risks effectively. Despite their differences in dimensions, most of these frameworks follow a similar path based on four main aspects: (1) identifying the risk sources for the supply chain; (2) assessing the supply chain risk probability and adverse consequences; (3) mitigating risks for the supply chain; and (4) monitoring the identified risks and the performance of risk mitigation activities (Jüttner, Peck, & Christopher, 2003; Rangel, de Oliveira, & Leite, 2015). Although extensive researches have been carried out on each of these areas, there are two major concerns regarding their applicability for risk management in biomass supply chain design and planning problems. First, given the fact that risk sources and their associated mitigating strategies are highly dependent on the type of supply chain and its environment, the existing frameworks that have been developed for general supply chains and not specifically for biomass supply chains cannot be directly applied to cope with all risks that may directly or indirectly influence biomass supply chain operations. Second, most previous studies focus on the qualitative understanding of supply chain risks while supply chain design and planning problems are interwoven with optimization approaches, highlighting the need for more quantitative insights in this field. To address these problems, this chapter designs a risk management framework for biomass supply chains that captures various dimensions of risk management to enable researchers to cope with a wide range of risks in a systematic way. Fig. 4.1 shows a schematic presentation of the proposed framework which consists of three key stages: (1) risk identification; (2) risk assessment; and (3) risk treatment. These stages are described in detail in the following sections.

4.2 Risk identification

The identification of risk sources is an important initial stage in the risk management process and its output is regarded as a foundation stone upon which other stages are built. As illustrated in Fig. 4.1, this stage aims to convert the vague and incomplete information of triggering events threatening the normal functioning of the supply chain into a clear and comprehensive vision of risk sources which allows decision makers to identify, classify, and locate any and all risks. Despite the importance of this stage, few studies have been conducted to identify the risks biomass supply chains are exposed to. Among them, Awudu and Zhang (2012) classified uncertainties

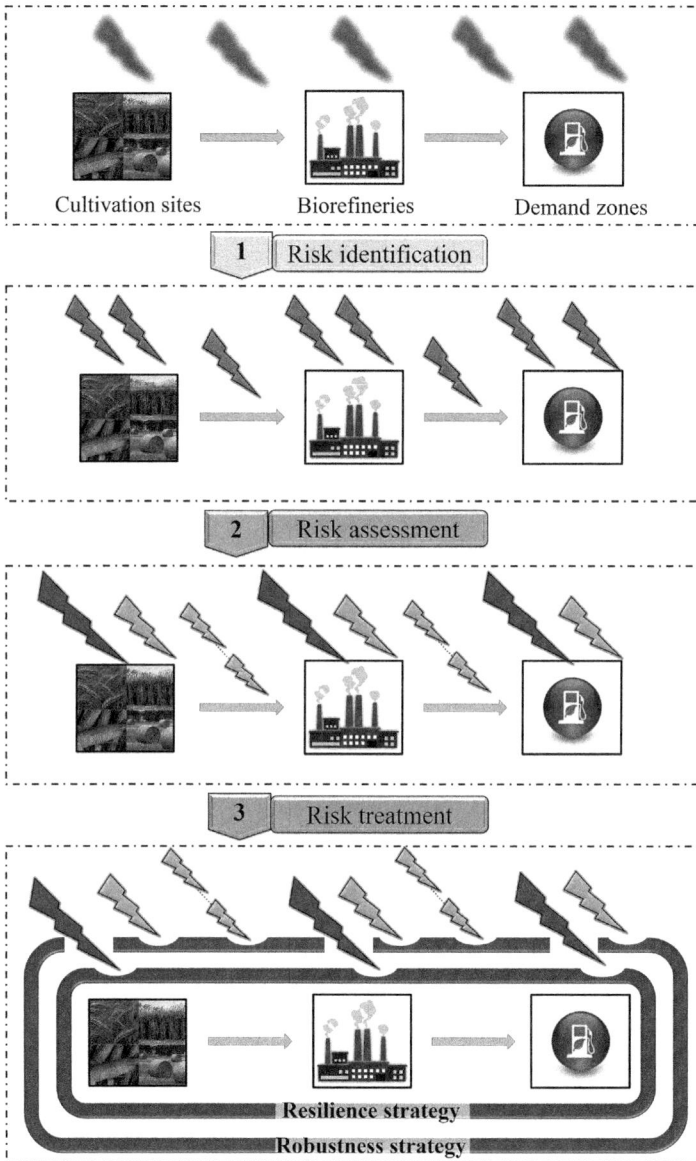

Figure 4.1 *A risk management framework for biomass-to-biofuel supply chains.*

affecting the performance of biomass supply chains into five categories: (1) biomass supply uncertainties; (2) transportation and logistics uncertainties; (3) production and operational uncertainties; (4) demand and price uncertainties; and (5) environmental uncertainties which include sustainability, tax, governmental and regulatory policies. Another taxonomy was proposed by Yatim, Lin, Lam, and Choy (2017), which groups main risks associated with the biomass industry in Malaysia into six classes: (1) supply chain and feedstock risks; (2) business risks; (3) technology risks; (4) financing risks; (5) regulatory risks; and (6) social and environmental risks. Depending on the time horizon of their impacts, Garcia and You (2015) suggested that uncertainties can fall under the two headings of strategic (long-term) and operational (short-term) uncertainties. The former involves uncertainties that remain unchanged for a specific period of time after realization such as changes in government incentives and policies, climate and weather effects, technology evolution and network stability, while the latter refers to those that change more frequently such as price and cost volatility, process yield, transportation and supply delays (Yue and You, 2016). In order to contribute to this growing area of research, a detailed typology of biomass supply chain risks is proposed in the following, based on an in-depth review of the existing literature on biomass supply chains and other related fields. Depending on their scope, controllability, predictability, and propagation properties, supply chain risks are classified by different typology schemes. A widely used scheme is to classify risks into three categories according to where in the supply chain they appear: (1) internal to the supply chain entities; (2) external to the supply chain entities but internal to the supply chain; and (3) external to the supply chain (Christopher and Peck, 2004). Inspired by this scheme, the risk sources of biomass supply chains are identified and classified into three categories through a typology presented in Fig. 4.2. These categories are further discussed in the following sections.

4.2.1 Internal risks to the supply chain entities

Internal risks to the supply chain entities refer to those that lie within the boundary of different entities of the biomass supply chain including biomass cultivation facilities, storage units, and biorefineries. The main types of internal risks are (1) technology performance uncertainty; (2) capital and operating cost uncertainty; (3) environmental and social impacts uncertainty; (4) yield uncertainty; and (5) raw material uncertainty.

- *Technology performance uncertainty:* Technology uncertainty stems from concerns that there is no guarantee that biofuel technologies operate and function as efficiently and effectively as expected and refers to issues such as unproven technology, engineering failure, and poor technology implementation (Yatim et al., 2017). The importance of this type of risk has been increasingly highlighted in recent years because the biofuel

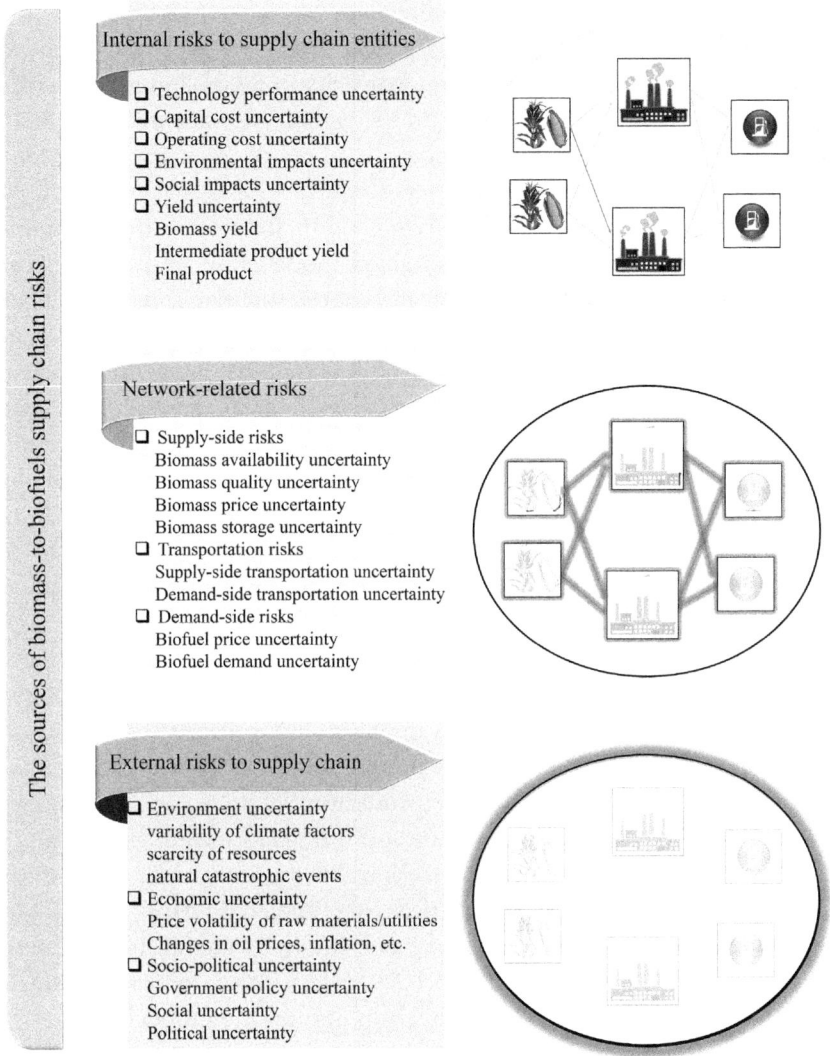

Internal risks to supply chain entities

❑ Technology performance uncertainty
❑ Capital cost uncertainty
❑ Operating cost uncertainty
❑ Environmental impacts uncertainty
❑ Social impacts uncertainty
❑ Yield uncertainty
   Biomass yield
   Intermediate product yield
   Final product

Network-related risks

❑ Supply-side risks
   Biomass availability uncertainty
   Biomass quality uncertainty
   Biomass price uncertainty
   Biomass storage uncertainty
❑ Transportation risks
   Supply-side transportation uncertainty
   Demand-side transportation uncertainty
❑ Demand-side risks
   Biofuel price uncertainty
   Biofuel demand uncertainty

External risks to supply chain

❑ Environment uncertainty
   variability of climate factors
   scarcity of resources
   natural catastrophic events
❑ Economic uncertainty
   Price volatility of raw materials/utilities
   Changes in oil prices, inflation, etc.
❑ Socio-political uncertainty
   Government policy uncertainty
   Social uncertainty
   Political uncertainty

The sources of biomass-to-biofuels supply chain risks

Figure 4.2 *Risks and uncertainties in biomass-to-biofuel supply chains.*

industry is moving from first-generation biofuel technology to second- and third-generation technologies which are still in development and not matured.

- *Capital and operating cost uncertainty:* The immaturity of processing technologies and lack of commercial-scale biofuel production have led to high variability in the costs associated with cultivation, harvesting, conversion and other production stages. For example, a harmonized cost comparison study examining differences in the reported production costs of microalgae-based biofuel reveals that the capital costs of an integrated biorefinery with an annual capacity of 50 million gallons range from $675.9 to $1332 million and its operating costs range from $216 to $383.5 million, based on a normalized set of input parameters and assumptions (Sun et al., 2011).

- *Environmental and social impacts uncertainty:* In addition to the economic dimension, some biomass supply chain models take into account the environmental and social impacts of biomass supply chains at the same time using a multiobjective optimization framework. The environmental and social impacts of the internal operations of the supply chain entities, which constitute a large part of the total impacts, have been analyzed in many previous studies, but there is high variability in their results. As a case in point, considering a similar proceeding method, the environmental impacts (GHG emissions) of biodiesel production from microalgae range from 200 g $CO_{2\text{-eq}}$ MJ^{-1} reported by Sills et al., (2012) to -41 $CO_{2\text{-eq}}$ MJ^{-1} reported by Quinn, Smith, Downes, and Quinn (2014) with a variety of values reported between these two extremes (Quinn and Davis, 2015). The high uncertainty in these results is brought about mainly by the limited knowledge about the production processes, different system boundaries, and unreliable input data. While most environmental impacts can be quantified by means of life cycle assessment (LCA) methods such as Eco-indicator 99, there is no quantitative evaluation method yet to measure social impacts including changes in social wellness, employment, personal rights, property land rights, health, food security, culture, labor practices, etc. (Yue et al., 2014). The lack of quantitative methods, on the one hand, and the wide diversity of social impacts, on the other hand, result in uncertainty in social parameters incorporated into biomass supply chain optimization models.

- *Yield and raw material uncertainty:* Another frequently observed type of risks in the biomass supply chain is high uncertainty in the yield of biomass in cultivation sites, in the yield of intermediate products from biomass in preprocessing facilities and in the yield of final products from intermediate products in biorefineries (Pérez et al., 2017). The yield of intermediate and final products can be estimated for first-generation biofuel supply chains that are working in many parts of the worlds because of availability of historical industrial data, while their estimation for a large number of second- and third-generation biomass feedstocks that have not been utilized for biofuel production at industrial scale is coupled with a high degree of uncertainty. The yield of biomass is more uncertain than that of intermediate and final products due to its dependence on climate factors (precipitation, sunlight, temperature, etc.) as well as lack of historical data for many energy crops that have not yet been commercially grown. A notable example is the biomass yield of microalgae whose estimated values vary by a factor of 6 from 8.2 to 50 $(g/m^2/d)$ (Tu, Eckelman, & Zimmerman, 2017). This can be contributed to the fact the yield values are estimated from the extrapolation of laboratory data obtained under highly controlled or/and short-time conditions which are not fully representative of large-scale operation (Quinn and Davis, 2015). In such uncertain situations, there is high variability in the amount of different raw materials required along the stages of the production chain, such as water, nutrients, and industrial chemicals. For instance, the amount of water consumed for jatropha cultivation is estimated between 60.34 and 70.64 kg based on 1 kg of produced biodiesel (Ajayebi, Gnansounou, & Raman, 2013) and is between 607 and 1944 kg for microalgae cultivation that is more water-intensive (Quinn and Davis, 2015), indicating that it is difficult or even impossible to determine the exact amount of their water consumption.

4.2.2 Network-related risks

Network-related risks, which are outside the supply chain entities but within the supply chain, stem from interactions along the chain (Jüttner et al., 2003). Depending on whether the risks reside in upstream or downstream operations, network-related risks are classified into supply- and demand-side risks. In the context of biomass supply chains, supply-side risks are related to biomass cultivation, harvesting, storage and transportation and mainly include biomass availability uncertainty, biomass quality uncertainty, biomass price uncertainty, biomass storage uncertainty and supply-side transportation uncertainty. On the other hand, demand-side risks are related to the transportation, distribution and sale of biofuels and mainly include biofuel demand uncertainty, biofuel price uncertainty and demand-side transportation uncertainty. These uncertainty types will be discussed in the following paragraphs.

- *Biomass availability, quality and price uncertainty:* The amount of biomass available at cultivation sites to be transported to biorefineries is associated with high level of uncertainty due to three main reasons: first, as stated earlier, biomass yield is influenced by weather and climate conditions, making biomass production at each site fluctuate from month to month and even from year to year. Second, cultivation sites are independent entities in nature which peruse their own goals and plans rather than those of the supply chain (Ye, Hou, Li, & Fu, 2018a). Their fragmented ownership and management can exacerbate the uncertainty related to biomass availability. Third, some types of biomass feedstocks such as forest residues are distributed in areas with limited accessibility during cold months, which can have a negative effect on biomass supply. Biomass quality plays a central role in determining the energy content of biomass and the amount of biomass required to produce one unit of biofuel and, consequently, is an influencing factor on the production costs of biofuel and the economic performance of biomass supply chains (Pérez et al., 2017). Biomass characteristics including density, moisture content, heating value, and chemical composition are heavily dependent on external factors such as soil type and fertilizers, growth conditions, climate, and water availability (Malladi and Sowlati, 2018), which vary from one cultivation site to another. Biomass collection from geographically dispersed sites, therefore, contributes to high variations in biomass quality. Biomass price is regarded as another important uncertain parameter in biomass supply chains because its small variations can noticeably change the cost of biofuel production (Shabani and Sowlati, 2016). Compared to the price of petroleum products that is mostly governed by supply and demand, the price of biomass is affected by several factors. A common factor between all biomass types is uncertainty in climate parameters and biomass quality, causing high variations in biomass prices during different supply periods. Another factor that is more related to first-generation biomass is price volatility in agricultural markets. Competition for edible biomass resources (e.g., sugar and oil) and agricultural inputs (e.g., water, fertilizer, and land) leads to a tightened interdependence and volatility spillover between bioenergy and food markets, making it more difficult and complicated to analyze and predict the price of biomass, mainly because of the noncooperative behavior of the parties involved. Nonedible feedstocks used to produce second- and third-generation biofuels are also influenced by this factor but to a lesser extent because they are able to grow on previously unused or marginal lands unsuitable for conventional agriculture with low or no consumption of agricultural water and fertilizer.

• *Biomass storage uncertainty:* Biomass supply uncertainty creates a need for biomass storage to ensure the continuous supply of biomass and match supply with demand. There are three locations for biomass storage, including supply fields, intermediate storage facilities (between supply sites and biorefineries), and biorefineries (Malladi and Sowlati, 2018). On-field storage that is generally done by ambient and covered storage methods is more convenient, time-saving and economical compared to other two techniques but suffers from many drawbacks, among which are lack of control over biomass moisture content, significant biomass material loss and high probability of fungus and spore formation (Rentizelas, Tolis, & Tatsiopoulos, 2009). Biomass storage at intermediate facilities and biorefineries is accompanied by densification processes with thermal and chemical treatments to decrease moisture content, increase the biomass to biofuels conversion rates and remove contaminants, enabling the prolonged storage of large volumes of biomass (Yue et al., 2014). Therefore, biomass degradation is a potential risk associated with the three modes of biomass storage but is more marked in the on-field storage than others due to its uncontrolled conditions. In biomass supply chain optimization problems, this risk is quantified by parameters such as dry matter losses and energy content reduction, which are highly uncertain and difficult to estimate because they are dependent on a large number of factors including the type of biomass and its properties, the location of storage facilities and their surrounding climate, the time and duration of storage, the method of storage and many others.

- *Supply- and demand-side transportation uncertainty:* Biomass supply chain transportation can be divided into two parts: (1) supply-side transportation for moving biomass between supply sites, intermediate storage facilities, and biorefineries and (2) demand-side transportation for moving biofuel between biorefineries and demand zones. The performance of these two parts can be affected by different problems, among which are road network congestion and accidents, tight pick-up and delivery time windows, loading and unloading restrictions, incompatibility of biomass and biofuel with transportation infrastructures, fuel price volatility, and poor communication and coordination (Sanchez-Rodrigues, Potter, & Naim, 2010). Incorporating all these potential risks into biomass supply chain design models, though appealing, can result in large-scale optimization problems that are computationally intractable. To remedy this difficulty, most supply chain models simplify the handling of them by only considering uncertainty in transportation cost and time with justification that the cumulative effects of transportation risks mainly manifest themselves in increased transportation cost and time. There is a marked difference between supply- and demand-side transportation in terms of cost and time. Biomass is bulky, prone to decomposition and widely geographically dispersed, making its transportation more costly and time-consuming than that of biofuel. Moreover, biomass transportation costs are subject to significant variability and difficult to estimate while biofuel transportation costs can often be estimated based on the transportation costs of its petroleum counterparts. The transportation costs of corn stover bales from the field to the biorefinery, for instance, are likely to vary in the range 35–50 ($/ton) depending on the corn stover harvest rate, dry matter loss, bale mass, number of bales per truck, lease cost of a truck and loader, biorefinery capacity, farmers' participation, and many others (Baral, Quiroz-Arita, & Bradley, 2017). Another type of uncertainty in biomass transportation is related to environmental and social impacts. Over the last years, significant efforts have been made to evaluate the environmental impacts of biomass and biofuel transportation. However, there is high variability in their results mainly due to using different system boundaries and different background data. Therefore, it is recommended to consider the environmental impacts of transportation as uncertain parameters in biomass supply chain optimization problems. The social impacts of biomass and biofuel transportation are unclear at the current stage of biofuel production. But, large-scale biofuel production that will probably occur in the near future leads to social impacts such as extra pressure on transportation services, more traffic congestion and accidents, higher risk of air pollution–related health effects and decreasing the beauties of the countryside (Mafakheri and Nasiri, 2014). These impacts are difficult to quantify and evaluate accurately, making uncertainty unavoidable in the evaluation of social impacts.

- *Biofuel price and demand uncertainty*: Biofuel price uncertainty as one of the downstream uncertainties can be attributed to two main reasons. First, the biofuel price uncertainty is affected by the upstream uncertainties given the fact that the uncertainty propagates through the supply chain from weather and climate parameters to biomass yield and price, then to total production costs and ultimately to biofuel price. Second, there is a significant volatility (uncertainty) spillover between biofuel and fossil fuel prices since the price of crude oil tends to lead the price of biofuel. Over the last decade, several empirical studies have confirmed the relationship between the price of biomass, biofuel and crude oil. Dutta, Bouri, Junttila, & Uddin, (2018), for example, investigate the corn-oil-ethanol nexus using daily US price index data for ethanol, corn, and oil from 2011 to 2016 and indicate that both corn and crude oil price uncertainties substantially impact US ethanol prices. Another source of the downstream uncertainties is biofuel demand uncertainty that is different between various types of biofuels. Demand for first-generation biofuels which are produced at large scale and have a growing share in the energy market generally fluctuates around a rising trend under the influence of variations in fossil fuel demand. However, demand for second- and third-generation biofuels that have not been commercialized yet is highly uncertain and unpredictable as a result of lack of their market data, and therefore, it is difficult to determine a narrow range of their demand fluctuations. There are other factors which further complicate the prediction of current and future demand for biofuels, such as changing regulations and policies, different regional economic structures and consumption behavior, increasing competitive pressure and market complexities, and the growth of hybrid and electric vehicles (Kim, Realff, & Lee, 2011; Yue et al., 2014).

4.2.3 External risks to the supply chain

External risks are defined as any uncertainty that may arise from outside the supply chain (Jüttner et al., 2003). Although beyond the realm of control of the supply chain, external risks must be addressed and handled by supply chain optimization models because they can directly or indirectly affect the operations of the supply chain. External risks to biomass supply chains are diverse but can be categorized into three groups: (1) environment uncertainty, (2) economic uncertainty, and (3) socio–political uncertainty, which will be described in the following paragraphs.

- *Environment uncertainty*: Biomass supply chains are exposed to four major types of environment uncertainty: variability of weather and climate factors (e.g., temperature, solar radiation, and precipitation), scarcity of resources (e.g., water and land), natural catastrophic events (e.g., floods, earthquakes, hurricanes, landslides, and droughts), and biotic stresses (e.g., pests and diseases). These uncertainty sources threaten the smooth operation among the various tiers of the supply chain and, if not managed properly, hinder the production of biofuel in a cost-effective and reliable manner because of several reasons such as biomass production reduction and subsequent increase in biomass prices due to the adverse effects of weather and climate variability, interruption of the supply chain operations due to damage to the supply chain facilities (e.g., cultivation sites, storage units, biorefineries, and distribution centers) and supporting infrastructures (e.g., electricity, water, transportation, and telecommunications), and volatility in biofuel prices due to variations in biomass and biofuel production (Langholtz et al., 2014).

- *Economic uncertainty*: One of the most recognized sources of economic uncertainty is the price volatility of raw materials (e.g., water, fertilizers, and chemicals) and utilities (e.g., electricity and gas). The reason behind the importance of these uncertainty sources is that the costs of raw materials and utilities constitute a significant portion of the operating costs of biofuel production and their small variations have a marked effect on the total operating costs. For example, an economic evaluation of wheat straw-to-butanol conversion based on a plant capacity of 150 million kg butanol/year indicates that the costs of raw materials and utilities (excluding the costs incurred for the purchase of wheat straw) account for 64% of annual operating costs, and a 20% increase in the prices of raw materials and utilities causes the total operating costs to grow by 12.7% from $193.9 to $218.6 million (Qureshi et al., 2013). Biomass supply chains are also exposed to other types of economic uncertainty, including changes in crude oil prices, inflation, interest rates, tax rates, minimum wage, exchange rates, and other macroeconomic variables.

- *Socio-political uncertainty*: This uncertainty category includes all possible threatening factors or actions emanating from any governmental body, political authority or social group and is divided into government policy uncertainty, social uncertainty, and political uncertainty (Bouchet, Clark, & Groslambert, 2003). Many governments around the world have promulgated supporting policies such as tax exemptions, subsidies and compulsory consumption schemes to promote renewable energy development, in general, and biofuels, in particular, which can facilitate the transition toward a more sustainable and secure energy future (Pérez et al., 2017). However, these policies are likely to be changed or even revoked after the establishment of biomass supply chains due to economic and environmental pressures. Changes or ambiguities in the regulations or policies related to the production of biofuel result in a discrepancy between realized and predicted returns, reducing the attractiveness of the biofuel industry to investors and other financing sources. These uncertainties also affect the performance of biomass supply chains. Therefore, ignoring the regulation- and policy-related uncertainties in biomass supply chain optimization models, particularly those with strategic long-term decisions, contributes to the unreliability of their results. Social uncertainty arises from negative public perceptions of impacts the biofuel industry has on communities and environments. The development of biofuels brings various social benefits such as employment creation, decrease in rural–urban migration, poverty reduction, rural development, and self-sufficiency in energy production (Mafakheri & Nasiri, 2014). However, there are numerous socio-environmental issues with biofuels such as diverting land, water, and other resources from food production to biofuel production and subsequent concerns about food security. Due to trade-offs between the advantages and disadvantages of biofuels, people in different regions have different attitudes towards biofuel production. Although this issue does not directly impact on the operational performance of biomass supply chains, it further complicates the estimation of biofuels demand at different locations because the demand at each location is dependent on the acceptance of biofuels in that location. The last type of socio-political uncertainty concerns political events or disruptions which influence the development and operations of biomass supply chains and includes terrorism, civil and international wars, riots, coups, insurrection, and sanctions.

4.3 Risk assessment

Having identified the uncertainties associated with biomass supply chains, the next stage is to assess them according to their probability to occur and associated consequences if they materialize (Sodhi and Tang, 2010). As indicated in Fig. 4.1, this stage evaluates each uncertainty identified in the identification stage and determines those that are more important to address. The outcome of the risk assessment can be used to provide a prioritized list of the uncertainties which helps decision makers identify those uncertainties with more severe consequences and those that are more likely to occur. In terms of their probability and consequences, biomass supply chain risks can be broadly categorized into two classes, namely operational risks and disruption risks. Operational risks, also called business-as-usual risks, refer to inherent uncertainties originating from high probability-low consequence events that frequently occur in the supply chain; for example, biomass supply and biofuel demand fluctuations (Sodhi, Son, & Tang, 2012). Although the negative impact of operational risks on the supply chain is usually considered to be of little importance, they can have serious consequences if they occur concurrently or trigger a snowball effect (Vilko & Hallikas, 2012). Disruption risks, on the other hand, are low probability–high consequence events stemming from natural and man-made disasters such as droughts and fuel price protests (Knemeyer, Zinn, & Eroglu, 2009).

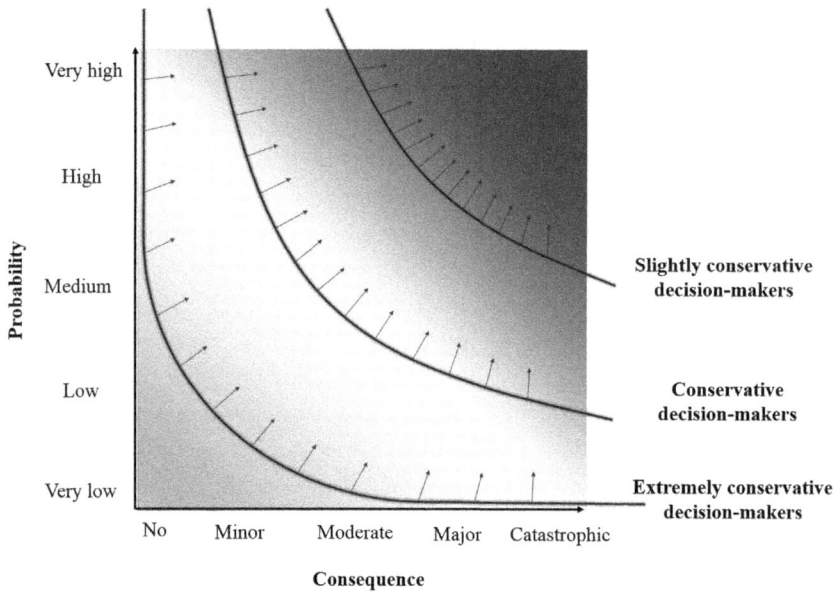

Figure 4.3 *Risk assessment matrix with different risk preferences.*

Such risks are seen as infrequent events; however, their effects may be so serious that the functionality of the supply chain is disrupted locally or globally for an unknown amount of time. It is obvious that between the operational and disruption risks as two extremes of potential risks there are many levels of risk with different probability and consequence measures. The most common tool to describe various levels of risk is the risk assessment matrix. This matrix plots supply chain risks on a graph with axes of probability and consequence as shown in Fig. 4.3. The two axes can be expressed in subjective terms (i.e., the Likert-type scales) or in objective terms (i.e., absolute values); the former is used for qualitative risk analyses and the latter for quantitative risk analyses (Waters, 2011). In the quantitative version, the risk matrix assigns a risk score to each risk, which is obtained by multiplying its probability by its consequence. In practice, however, neither the probability nor the consequence can be accurately estimated as a numerical value. The qualitative version categorizes the probability and the consequence into a number of levels based on the Likert-type scale and then scores each level, usually on a scale of 1–5, using expert opinion. This process produces a matrix with multiple cells, each representing a level of risk. The score of each level is calculated by multiplying its corresponding scores of the probability and the consequences together. With the help of the obtained scores, a most-to-least critical risk ranking can be produced. Using the risk matrix, all the risks facing biomass supply chains can be evaluated and visualized. It should be noted that one must not expect to find a similar risk matrix for different biomass supply chains because the probability and the consequence are strongly dependent on where the supply chain is established and its surrounding risks. After the risk matrix is formed, the acceptable level of risk must be defined in order to determine which risks can be accepted or ignored and which risks need treatment. The rationale behind this differentiation is that the cost of mitigating some risks is higher than the cost of bearing them, and decision makers believe that it is not economical to develop strategies for combating such risks (Scholten and Fynes, 2017). Moreover, it is not possible in most supply chain optimization problems to manage all the identified risks mainly due to time and computational limitations. Since risk is often viewed as a subjective concept that is highly related to the individual's assessment of the importance of risk, the risk attitude of decision makers plays an important role in determining the acceptable level (Heckmann, Comes, & Nickel, 2015). The risk attitude can be incorporated in the risk assessment by defining three types of risk preferences: slightly conservative, conservative, and extremely conservative. As illustrated in Fig. 4.3, slightly conservative decision-makers focus only on risks whose probability and consequence are high and perceive other risks as acceptable. Conservative decision-makers are sensitive to a wider range of risks that includes risks whose probability or consequence, or both, are high or moderate. Extremely conservative decision-makers consider all potential risks, except low-probability, low-consequence risks, as serious risks that must be involved in the risk management process.

4.4 Risk treatment

After assessing the identified risks and choosing those that must be dealt with, the next step is to develop suitable approaches for managing the risks. Risk management strategies applied in the problems of supply chain design and planning under uncertainty can be classified into two broad categories: robustness and resilience enhancement approaches. As indicated in Fig. 4.1, robustness approaches protect the supply chain against high probability-low consequence risks (i.e., business-as-usual risks) and resilience approaches protect the supply chain against low probability-high consequence risks (i.e., disruption risks) (Behzadi et al., 2018). The effect of these two approaches on the performance of the supply chain is illustrated in Fig. 4.4.

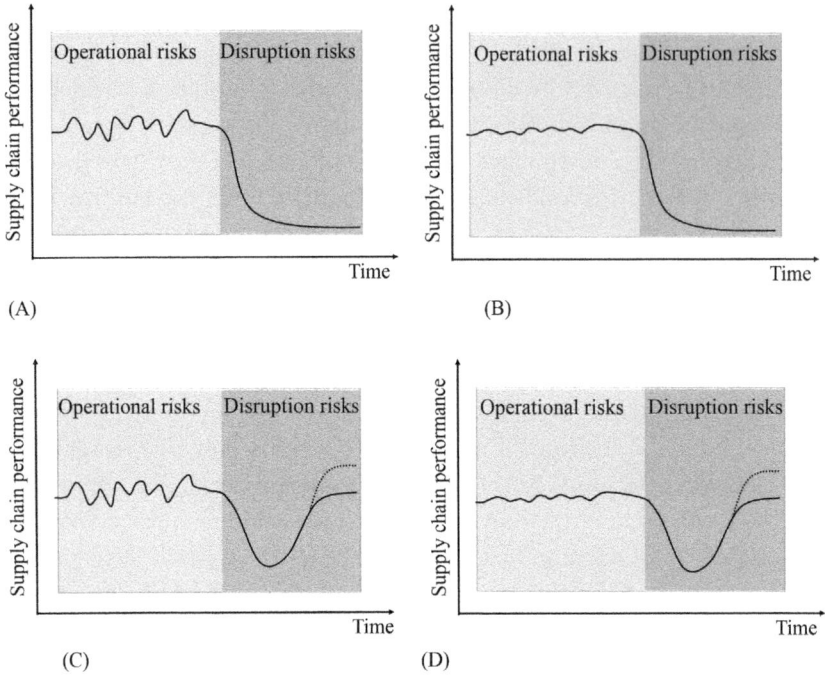

Figure 4.4 *The performance of supply chain with and without robustness and resilience enhancement approaches.* (A) Without robustness and resilience approaches, (B) with only robustness approaches, (C) with only resilience approaches, and (D) with both robustness and resilience approaches.

Robustness approaches enable the supply chain to withstand operational fluctuations, retain its original structure and stay functional in the face of business-as-usual risks, thereby ensuring a steady performance (Behzadi et al., 2017). Resilience approaches, on the other hand, allow the supply chain to return quickly and smoothly to its original state or even move to a more desirable state after being disturbed (Christopher & Peck, 2004). Therefore, for effective supply chain risk management, both robustness and resilience approaches must be considered in the modeling and optimization of biomass supply chains.

4.4.1 Robustness enhancement approaches

Deterministic biomass supply chain design and planning models, which ignore uncertainties, result in solutions (i.e., supply chain decisions) that may be suboptimal, infeasible or both in the presence of uncertainty (Shabani and Sowlati, 2016). In contrast, nondeterministic models aim to

find solutions that can be trusted to perform well under possible realizations of uncertainties (Govindan, Fattahi, & Keyvanshokooh, 2017). There
are several nondeterministic optimization approaches with different definitions and measurements of good performance as well as with different
levels of robustness enhancement. In choosing the most suitable approach
for handling uncertainty in supply chains, one must consider the amount of
data available to model uncertainty. The importance of this criterion lies in
the fact that the amount of available data is closely related to the type (i.e.,
nature) of uncertainty and its corresponding treatment approaches. When
uncertainty, specifically in the context of supply chain optimization, is examined through the lens of availability of data, three types of uncertainty
can be distinguished: randomness uncertainty, epistemic uncertainty, and
deep uncertainty (Bairamzadeh, Saidi-Mehrabad, & Pishvaee, 2018). Under
randomness uncertainty, there is reliable historical data to exactly or approximately estimate the probability distributions quantifying the uncertain
parameters of supply chain models (Klibi, Martel, & Guitouni, 2010). The
uncertainty in the demand of commonly used biofuels, for example, belongs to this type of uncertainty because its probability distributions can be
perfectly or ambiguously obtained with the help of available data. However,
it is impossible in many real-world problems to extract probability distributions due to limited or unreliable data (Pishvaee, Rabbani, & Torabi, 2011).
In such situations, subjective probability distributions can be estimated
based on experts' judgments. When the opinion of experts is elicited, the
epistemic uncertainty arises because of the lack of knowledge about the
credibility of information provided by them (Hester, 2012). The estimates
of biomass quality derived from expert knowledge, for example, are tainted
with epistemic uncertainty. Under deep uncertainty, the available data are
so limited that an objective or subjective probability distribution cannot be
obtained, and only the bounds of uncertain parameters can be estimated
(Klibi et al., 2010). The uncertainty in the yield of nonedible energy crops
not previously grown on a commercial scale comes to mind as an example
of deep uncertainty. After determining the type of uncertainty, an appropriate optimization approach for coping with it must be selected. Fig. 4.5 provides an overview of leading optimization approaches for hedging against
various types of uncertainty, which are described in the following.

- *Stochastic programming*: As a conventional approach to tackle the randomness uncertainty, stochastic programming treats uncertain parameters as
  random variables following a known probability distribution. Since employing continuous distribution functions in real-world complex supply

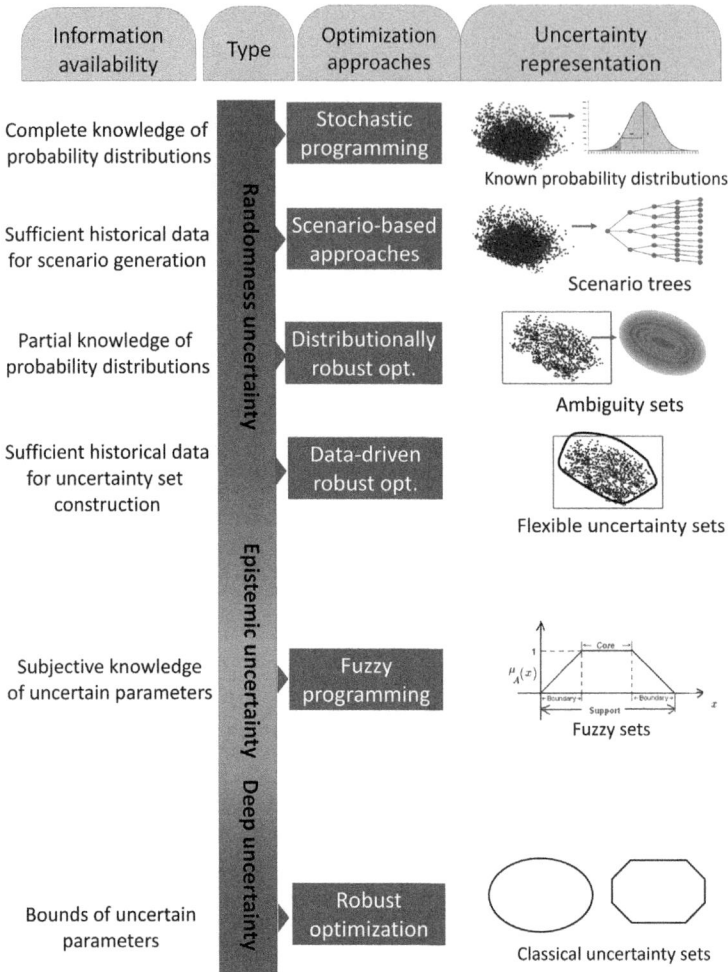

Figure 4.5 *An overview of leading optimization approaches and their characteristics.*

chains leads to computationally intractable problems, uncertain parameters are often described by a set of discrete scenarios, each with a probability of occurrence (Govindan et al., 2017). In general, supply chain
problems are formulated as two-stage stochastic programming where
first-stage decisions corresponding to strategic (long-term) decisions
such as the selection of production technologies must be made before
the uncertainty is realized and second-stage decisions corresponding to
tactical/operational (mid-term/short-term) decisions such as the transportation of biomass and biofuels can be made after the values of the

uncertain parameters are disclosed. To better incorporate sequences of decisions over time, two-stage stochastic programming can be extended to a multistage setting where the uncertainty is revealed sequentially, enabling the decision maker to take corrective actions after some uncertainty is resolved and more information is obtained (Gupta and Grossmann, 2011). Despite its strong theoretical foundation, stochastic programming has three main limitations in practice. First, in view of the fact that the optimal solution of scenario-based stochastic programming approaches is feasible only for the considered scenarios, a slight deviation from the input scenarios, which is very likely to occur, renders the solution suboptimal or even infeasible. Second, as the number of uncertain parameters increases, the number of scenarios and the computational complexity of the problem grow exponentially, resulting in intractable problems (Mohseni and Pishvaee, 2016). Third, typical scenario-based stochastic models optimize the expected value of the objective function, yielding a solution that performs well on average without ensuring that it is immunized against worst-case scenarios (Zokaee et al., 2017).

- *Robust optimization*: The main idea of robust optimization is to find a robust solution using worst-case analysis. A solution is called robust if it is the best optimal solution that is guaranteed to remain feasible for all possible realizations of the uncertain parameters. In order to avoid the possibility of infeasibility, robust optimization approaches focus on optimization over the worst-case realizations. Therefore, robust solutions are less sensitive or even insensitive to fluctuations in the uncertain parameters within a specific range (Alvarez and Vera, 2014). This feature is particularly useful in immunizing strategic level decisions such as the location of biorefineries because these decisions often remain unchanged for years or decades once they are made (Garcia, Gong, & You, 2016). Depending on the way the uncertainty is represented, robust optimization approaches can be divided into two main classes: scenario-based robust optimization and uncertainty set-based robust optimization. The former, which uses a number of discrete scenarios for describing randomness uncertainty, is able to control the degree of robustness and feasibility but may have difficulty in handling a large number of scenarios and in constructing scenarios due to limited historical data (Pishvaee, Razmi, & Torabi, 2012). The latter, which has emerged as a powerful tool for optimization under deep uncertainty, assumes that possible realizations lie within a continuous uncertainty set. Robust optimization has received increasing attention in recent years because it only requires the lower and upper bounds of the uncertain parameters instead of detailed probabilistic knowledge and also circumvents the computational difficulties posed by scenario-based approaches (Mohseni et al., 2016). The first step in this direction was taken by Soyster (1973) who proposed a worst-case model so as to satisfy all constraints under possible realizations belonging to an interval set. Although the Soyster's approach ensures the highest protection, it tends to generate over conservative solutions which are not economically justifiable. Moreover, it is very unlikely, at least in most supply chain problems, that all uncertain parameters reach their worst values simultaneously. This problem has been overcome by the development of different types of uncertainty sets such as polyhedral and ellipsoidal sets that control the conservatism level of the solution, allowing a trade-off between the solution robustness and its cost (Li, Ding, & Floudas, 2011).

- *Distributionally robust optimization*: Although complete information about the uncertain parameters required in stochastic programming is seldom available in practice, there may be incomplete historical data such as a finite number of observed samples or partial distributional information of the uncertain parameters, which can be utilized to improve the reliability of the solutions provided by robust optimization. As an intermediate approach between stochastic programming and robust optimization, distributionally robust optimization (DRO) presumes that probability distributions governing the uncertain parameters are unknown and subject to uncertainty. In order to hedge against the ambiguity of probability distributions, DRO considers a family of probability distributions (called the ambiguity set) which is assumed to incorporate the true distribution (Rahimian & Mehrotra, 2019). The ambiguity set can be obtained even from partial historical data and makes DRO less sensitive to the deviations from the true distribution compared to stochastic programming. For the construction of the ambiguity set, a number of methods have been proposed (Esfahani & Kuhn, 2018; Shang & You, 2018), which employ different mechanisms to extract statistical information from available data, resulting in different DRO approaches with different levels of computational efficiency and conservatism (Duan et al., 2017). Generally, the idea of DRO is to optimize the worst-case expected performance over a set of probability distributions within the ambiguity set, while robust optimization aims to optimize the worst-case performance over an uncertainty set without considering any distributional information. In fact, if the ambiguity set contains distributions that place all their weight on the boundary of the uncertainty set, then DRO reduces to robust optimization (Delage & Ye, 2010).

- *Data-driven robust optimization*: The uncertainty set-induced robust optimization framework suffers from two important problems. First, the bounds of the uncertain parameters that determine the size of the uncertainty set are difficult to adjust, particularly when the number of uncertain parameters grows. In order to include every possible realization, a common way is to consider wide variation ranges for the uncertain parameters. However, this procedure yields highly conservative solutions in exchange for a significant sacrifice in the objective function value which is interpreted as the cost of robustness (Mohseni & Pishvaee, 2019). The second problem is associated with the geometric structure of the uncertainty set as another factor affecting the performance of robust optimization approaches. The conventional uncertainty sets are limited in number and their structures are built *a priori* without knowledge of the underlying uncertainty (Ning & You, 2019). These limitations make it difficult to find an uncertainty set that can capture the intrinsic characteristics of the uncertain parameters. Motivated by the above problems, data–driven robust optimization (DDRO) has emerged as a promising optimization framework for providing high-quality robust solutions when there is at least a set of data samples that can be a good representative of the whole uncertainty domain. Taking into account the correlations and asymmetry that may be embodied in the uncertainty data, DDRO constructs a compact uncertainty set that encloses the region in which data samples reside (Shang, Huang, & You, 2017). The flexible uncertainty set provided by DDRO encapsulates the most likely realizations of the uncertain parameters rather than every possible realization, thereby avoiding unnecessary conservatism associated with the classical uncertainty sets.
- *Fuzzy programming*: As an important branch of optimization under uncertainty, fuzzy mathematical programming can treat epistemic uncertainty in the setting of optimization problems. When human judgments and preferences are involved in the decision-making process, epistemic uncertainty arises due to the lack of complete knowledge about the parameters, constraints, and goals of the underlying optimization models (Naderi, Pishvaee, & Torabi, 2016). Using subjective knowledge and preferences of the decision maker, epistemic uncertainty is formulated in the form of fuzzy sets characterized by membership functions. Fuzzy mathematical programming is classified into two broad catego-

ries: flexible programming and possibilistic programming (Inuiguchi and Ramık, 2000). Flexible programming is utilized to deal with fuzzy goals and constraints stemming from the vague preference of the decision maker (Pishvaee & Khalaf, 2016). Fuzzy goals appear when the decision-maker sets a flexible target for the objective function instead of strictly optimizing the objective function, and fuzzy constraints represent the elasticity of soft constraints where a certain amount of constraint violation is allowed. Possibilistic programming handles the fuzziness of the parameters whose values are subjectively estimated by the decision-maker due to the unavailability or insufficiency of objective data (Mousazadeh, Torabi, & Pishvaee, 2014). In supply chain optimization problems, flexible and possibilistic programming are used independently on in combination to cope with various forms of fuzziness.

A detailed description of the above optimization approaches, including mathematical formulations, is given in Chapter 6, entitled "Uncertainty modeling approaches for biofuel supply chains".

4.4.2 Resilience enhancement approaches

A review of studies on supply chain optimization under disruption risks shows that there are two main ways to make the supply chain more resilient against disruptions. The first is using an optimization approach that often models disruptions as a set of discrete scenarios representing possible changes in supply chain components (e.g., suppliers, transporting links and production facilities) when a disruption occurs and then determines optimal supply chain decisions under the defined scenarios. The second utilizes resilience strategies with the purpose of reducing supply chain vulnerability. According to Tomlin (2006), resilience strategies are categorized into two groups: mitigation and contingency strategies. Mitigation strategies, also called pre-disruption or proactive strategies, refer to preventive actions taken in advance of disruptions and therefore incur a cost regardless of whether a disruption happens. However, contingency strategies, also called postdisruption or reactive strategies, are performed when a disruption takes place with the aim of returning the supply chain to its initial state. There are various resilience strategies to hedge against disruption risks and enhance the resiliency of biomass supply chains (see Fig. 4.6). These strategies can be broadly grouped into three classes which will be described in the following.

- *Resilience in biomass supply*: Most biomass supply chain models developed in the literature are restricted to the use of a single biomass feedstock cultivated in limited areas with similar climate and geographical

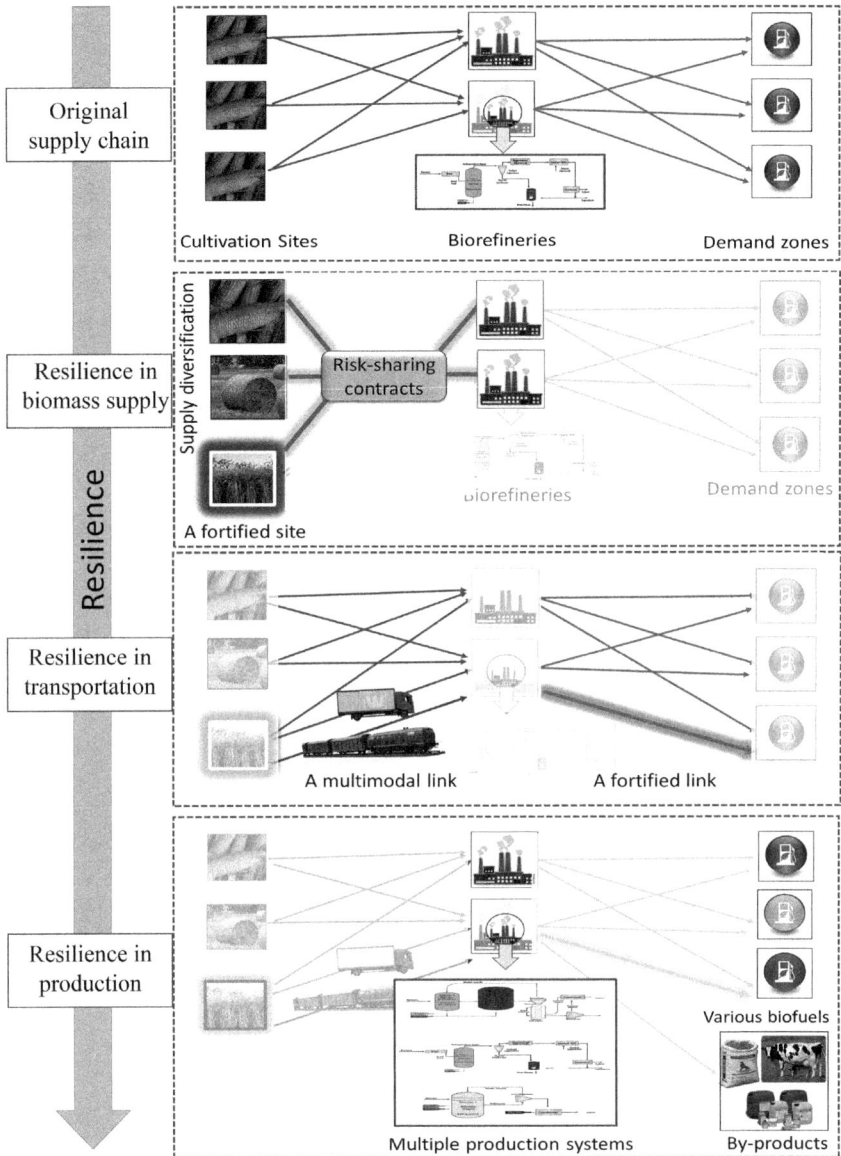

Figure 4.6 *An overview of resilience strategies to enhance the resiliency of biomass supply chains.*

characteristics. This issue significantly increases the risk of supply disruption because climate extremes such as drought, floods, and destructive storms can completely disrupt the supply of biomass and subsequently the operations of the supply chain. The cultivation of biomass in geographically dispersed locations can alleviate this concern since it has the advantage that not all cultivation sites are simultaneously affected by climate extremes, which avoids the complete disruption of biomass supply. Another strategy to enhance flexibility in biomass supply is to use several types of biomass feedstock. As different biomass feedstocks have different planting, maturation and harvesting periods as well as different sensitivities to climate change and environmental extremes (Langholtz et al., 2014), the diversification strategy not only improves the capability of the supply chain to maintain the supply of biomass when facing a disruption, but also helps to achieve a continuous and balanced supply throughout the year. In recent years, setting risk-sharing coordination contracts between farmers (i.e., biomass suppliers) and biofuel producers (i.e., biomass buyers) has been proposed as a resilience strategy to ensure the sustainable supply of biomass (Ye, Hou, Li, & Fu, 2018a). With such contracts, when there is a difference between the realized yield of farmers and the order quantity of biofuel producers, the risk of the over-production or under-production of biomass is shared among all parties (Ye, Li, & Yang, 2018b). The impact of disruption risks on biomass supply can also be mitigated through the fortification of cultivation sites by, for example, building floodwalls for protection against seasonal or extreme weather events.

- *Resilience in transportation*: As an experience from the 2003 U.S. Northeast blackout, Hurricane Katrina in 2005, Typhoon Morakot in Taiwan in 2009, and the earthquake and tsunami in Japan in 2011 shows, transportation infrastructures and logistics operations are highly vulnerable to natural and man-made disasters (Liao, Hu, & Ko, 2018; Marufuzzaman & Ekşioğlu, 2016). The main effect of such disasters on supply chain transportation is the failure of a number of transportation links which further complicates logistics management and increases transportation time and costs. Moreover, the seasonality of biomass makes biomass supply chains more susceptible to disruption because the harvest and transportation season of some biomass feedstocks (e.g., corn stover) coincides with the season of hurricanes and floods (Marufuzzaman & Ekşioğlu, 2016). There are various strategies to add more flexibility in transportation systems with the aim of protecting them against disruptions. Multimodal transportation is one of the effective strategies to prevent the regular operation of the supply chain from being halted when a disruption occurs in maritime shipping, air transportation, road and rail networks, etc., in view of the fact that it utilizes multiple modes of transportation rather than single mode of transportation, which ensures that transportation systems function well not only under normal conditions but also in the face of unexpected disruptions (Tang, 2006). Fortification of transportation links is another strategy that could be adopted (Poudel et al., 2016). This strategy suggests strengthening a number of links, particularly those connecting the vital players in the supply chain (e.g., main biomass suppliers and biorefineries) and those located closer to the risk areas of potential disasters, thereby decreasing the probability of failures and improving the overall reliability of the transportation network.

- *Resilience in production*: Many of the existing biorefineries employ a fixed production process that accepts a single biomass feedstock and produces one main product. Despite its simplicity, this production process is not reliable enough to maintain uninterrupted production in case of disruption events because the shortage of a particular feedstock can bring a biorefinery to a halt (Langholtz et al., 2014). To prevent a complete shutdown, biorefineries must be equipped with multiple production technologies capable of accommodating a broad range of feedstocks and yielding different types of biofuels and a wide variety of by-products (e.g., soil supplements, animal feed additives, and industrial chemicals) (Wang, Huo, & Arora, 2011). Although the diversification of production systems enables the supply chain to respond more adaptively to supply and demand disruptions, it increases the complexity in operations and production-related tasks which negatively impacts on the resilience of the supply chain (Pérez et al., 2017). Therefore, there is a need to investigate the trade-off between the positive and negative aspects of this strategy in terms of resilience before adopting it.

4.5 Conclusions

Biomass-to-biofuel supply chains are exposed to a wide range of uncertainties and risks arising from issues such as technology evolution, changing policies and regulations, demand and price variability, unpredictable weather conditions, production cost variations as well as man-made and natural disasters. Failure to hedge against all such uncertainties may result in suboptimal or even infeasible supply chain decisions. However, most previous studies on biofuel supply chain optimization under uncertainty deal with only one or a few types of uncertainties without any reasonable justification for doing so. This chapter proposes a systematic risk management framework for the biofuel supply chain, which consists of three main stages, namely risk identification, risk assessment, and risk treatment. The first stage provides a comprehensive list of all possible risks that may negatively affect the normal operation of the supply chain and classifies them into three categories: internal risks to supply chain entities, network-related risks and external risks to supply chain, helping the decision-maker become aware of where they may arise from. This stage is followed by the risk assessment that evaluates the identified risks based on their probability and consequence to determine those that must be addressed first. Finally, risk treatment presents risk management strategies along with a variety of prominent optimization approaches, aiming to enhance the resilience and robustness of the supply chain against operational and disruption risks.

References

Ajayebi, A., Gnansounou, E., & Raman, J. K. (2013). Comparative life cycle assessment of biodiesel from algae and jatropha: A case study of India. *Bioresource Technology*, *150*, 429–437.

Alvarez, P. P., & Vera, J. R. (2014). Application of robust optimization to the sawmill planning problem. *Annals of Operations Research*, *219*, 457–475.

Awudu, I., & Zhang, J. (2012). Uncertainties and sustainability concepts in biofuel supply chain management: A review. *Renewable and Sustainable Energy Reviews*, *16*, 1359–1368.

Bairamzadeh, S., Saidi-Mehrabad, M., & Pishvaee, M. S. (2018). Modelling different types of uncertainty in biofuel supply network design and planning: A robust optimization approach. *Renewable Energy*, *116*, 500–517.

Baral, N. R., Quiroz-Arita, C., & Bradley, T. H. (2017). Uncertainties in corn stover feedstock supply logistics cost and life-cycle greenhouse gas emissions for butanol production. *Applied Energy*, *208*, 1343–1356.

Behzadi, G., O'Sullivan, M. J., Olsen, T. L., & Zhang, A. (2018). Agribusiness supply chain risk management: A review of quantitative decision models. *Omega*, *79*, 21–42.

Behzadi, G., O'Sullivan, M. J., Olsen, T. L., Scrimgeour, F., & Zhang, A. (2017). Robust and resilient strategies for managing supply disruptions in an agribusiness supply chain. *International Journal of Production Economics*, *191*, 207–220.

Bouchet, M. H., Clark, E., & Groslambert, B. (2003). *Country risk assessment: A guide to global investment strategy*. England: John Wiley & Sons.

Christopher, M., & Peck, H. (2004). Building the resilient supply chain. *The International Journal of Logistics Management*, *15*, 1–14.

Delage, E., & Ye, Y. (2010). Distributionally robust optimization under moment uncertainty with application to data-driven problems. *Operations Research*, *58*, 595–612.

Duan, C., Jiang, L., Fang, W., & Liu, J. (2017). Data-driven affinely adjustable distributionally robust unit commitment. *IEEE Transactions on Power Systems*, *33*, 1385–1398.

Dutta, A., Bouri, E., Junttila, J., & Uddin, G. S. (2018). Does corn market uncertainty impact the US ethanol prices? *Gcb Bioenergy*, *10*, 683–693.

Esfahani, P. M., & Kuhn, D. (2018). Data-driven distributionally robust optimization using the Wasserstein metric: Performance guarantees and tractable reformulations. *Mathematical Programming*, *171*, 115–166.

Garcia, D. J., & You, F. (2015). Supply chain design and optimization: Challenges and opportunities. *Computers & Chemical Engineering*, *81*, 153–170.

Garcia, D. J., Gong, J., & You, F. (2016). Multi-stage adaptive robust optimization over bio-conversion product and process networks with uncertain feedstock price and biofuel demand. In *Computer Aided Chemical Engineering* (pp. 217–222). (38). Elsevier.

Govindan, K., Fattahi, M., & Keyvanshokooh, E. (2017). Supply chain network design under uncertainty: A comprehensive review and future research directions. *European Journal of Operational Research, 263*, 108–141.

Gupta, V., & Grossmann, I. E. (2011). Solution strategies for multistage stochastic programming with endogenous uncertainties. *Computers & Chemical Engineering, 35*, 2235–2247.

Heckmann, I., Comes, T., & Nickel, S. (2015). A critical review on supply chain risk–Definition, measure and modeling. *Omega, 52*, 119–132.

Hester, P. (2012). Epistemic uncertainty analysis: An approach using expert judgment and evidential credibility. *International Journal of Quality, Statistics, and Reliability, 2012*.

Inuiguchi, M., & Ramık, J. (2000). Possibilistic linear programming: A brief review of fuzzy mathematical programming and a comparison with stochastic programming in portfolio selection problem. *Fuzzy Sets and Systems, 111*, 3–28.

Jüttner, U., Peck, H., & Christopher, M. (2003). Supply chain risk management: Outlining an agenda for future research. *International Journal of Logistics: Research and Applications, 6*, 197–210.

Kim, J., Realff, M. J., & Lee, J. H. (2011). Optimal design and global sensitivity analysis of biomass supply chain networks for biofuels under uncertainty. *Computers & Chemical Engineering, 35*, 1738–1751.

Klibi, W., Martel, A., & Guitouni, A. (2010). The design of robust value-creating supply chain networks: A critical review. *European Journal of Operational Research, 203*, 283–293.

Knemeyer, A. M., Zinn, W., & Eroglu, C. (2009). Proactive planning for catastrophic events in supply chains. *Journal of Operations Management, 27*, 141–153.

Langholtz, M., Webb, E., Preston, B. L., Turhollow, A., Breuer, N., Eaton, L., & Downing, M. (2014). Climate risk management for the US cellulosic biofuels supply chain. *Climate Risk Management, 3*, 96–115.

Li, Z., Ding, R., & Floudas, C. A. (2011). A comparative theoretical and computational study on robust counterpart optimization: I. Robust linear optimization and robust mixed integer linear optimization. *Industrial & Engineering Chemistry Research, 50*, 10567–10603.

Liao, T.-Y., Hu, T.-Y., & Ko, Y.-N. (2018). A resilience optimization model for transportation networks under disasters. *Natural Hazards, 93*, 469–489.

Mafakheri, F., & Nasiri, F. (2014). Modeling of biomass-to-energy supply chain operations: Applications, challenges and research directions. *Energy Policy, 67*, 116–126.

Malladi, K. T., & Sowlati, T. (2018). Biomass logistics: A review of important features, optimization modeling and the new trends. *Renewable and Sustainable Energy Reviews, 94*, 587–599.

Marufuzzaman, M., & Eksˌiog˘lu, S. D. (2016). Designing a reliable and dynamic multimodal transportation network for biofuel supply chains. *Transportation Science, 51*, 494–517.

Mohseni, S., & Pishvaee, M. S. (2016). A robust programming approach towards design and optimization of microalgae-based biofuel supply chain. *Computers & Industrial Engineering, 100*, 58–71.

Mohseni, S., Pishvaee, M. S., & Sahebi, H. (2016). Robust design and planning of microalgae biomass-to-biodiesel supply chain: A case study in Iran. *Energy, 111*, 736–755.

Mohseni, S., & Pishvaee, M. S. (2019). Data-driven robust optimization for wastewater sludge-to-biodiesel supply chain design. *Computers & Industrial Engineering, 139*, 105944.

Mousazadeh, M., Torabi, S. A., & Pishvaee, M. S. (2014). Green and reverse logistics management under fuzziness. In *Supply chain management under fuzziness* (pp. 607–637). Springer.

Naderi, M. J., Pishvaee, M. S., & Torabi, S. A. (2016). Applications of fuzzy mathematical programming approaches in supply chain planning problems. In *Fuzzy Logic in Its 50th Year* (pp. 369–402). Springer.

Ning, C., & You, F. (2019). Optimization under uncertainty in the era of big data and deep learning: When machine learning meets mathematical programming. *Computers & Chemical Engineering, 125*, 434–448.

Pishvaee, M. S., Rabbani, M., & Torabi, S. A. (2011). A robust optimization approach to closed-loop supply chain network design under uncertainty. *Applied Mathematical Modelling, 35*, 637–649.

Pishvaee, M. S., & Khalaf, M. F. (2016). Novel robust fuzzy mathematical programming methods. *Applied Mathematical Modelling, 40*, 407–418.

Pishvaee, M. S., Razmi, J., & Torabi, S. A. (2012). Robust possibilistic programming for socially responsible supply chain network design: A new approach. *Fuzzy Sets and Systems, 206*, 1–20.

Poudel, S. R., Marufuzzaman, M., & Bian, L. (2016). Designing a reliable bio-fuel supply chain network considering link failure probabilities. *Computers & Industrial Engineering, 91*, 85–99.

Pérez, A. T. E., Camargo, M., Rincón, P. C. N., & Marchant, M. A. (2017). Key challenges and requirements for sustainable and industrialized biorefinery supply chain design and management: A bibliographic analysis. *Renewable and Sustainable Energy Reviews, 69*, 350–359.

Quinn, J. C., Smith, T. G., Downes, C. M., & Quinn, C. (2014). Microalgae to biofuels life-cycle assessment—Multiple pathway evaluation. *Algal Research, 4*, 116–122.

Quinn, J. C., & Davis, R. (2015). The potentials and challenges of algae based biofuels: A review of the techno-economic, life cycle, and resource assessment modeling. *Bioresource Technology, 184*, 444–452.

Qureshi, N., Saha, B., Cotta, M., & Singh, V. (2013). An economic evaluation of biological conversion of wheat straw to butanol: A biofuel. *Energy Conversion and Management, 65*, 456–462.

Rahimian, H., & Mehrotra, S. (2019). Distributionally robust optimization: A review. *ArXiv Preprint ArXiv:1908.05659*.

Rangel, D. A., de Oliveira, T. K., & Leite, M. S. A. (2015). Supply chain risk classification: Discussion and proposal. *International Journal of Production Research, 53*, 6868–6887.

Rentizelas, A. A., Tolis, A. J., & Tatsiopoulos, I. P. (2009). Logistics issues of biomass: The storage problem and the multi-biomass supply chain. *Renewable and Sustainable Energy Reviews, 13*, 887–894.

Sanchez-Rodrigues, V., Potter, A., & Naim, M. M. (2010). Evaluating the causes of uncertainty in logistics operations. *The International Journal of Logistics Management, 21*, 45–64.

Scholten, K., & Fynes, B. (2017). Risk and uncertainty management for sustainable supply chains. In *Sustainable supply chains* (pp. 413–436). Springer.

Shabani, N., & Sowlati, T. (2016). A hybrid multi-stage stochastic programming-robust optimization model for maximizing the supply chain of a forest-based biomass power plant considering uncertainties. *Journal of Cleaner Production, 112*, 3285–3293.

Shang, C., Huang, X., & You, F. (2017). Data-driven robust optimization based on kernel learning. *Computers & Chemical Engineering, 106*, 464–479.

Shang, C., & You, F. (2018). Distributionally robust optimization for planning and scheduling under uncertainty. *Computers & Chemical Engineering, 110*, 53–68.

Sills, D. L., Paramita, V., Franke, M. J., Johnson, M. C., Akabas, T. M., Greene, C. H., & Tester, J. W. (2012). Quantitative uncertainty analysis of life cycle assessment for algal biofuel production. *Environmental Science & Technology, 47*, 687–694.

Sodhi, M. S., & Tang, C. S. (2010). Supply chain risk management. In *Wiley encyclopedia of operations research and management science*.

Sodhi, M. S., Son, B., & Tang, C. S. (2012). Researchers' perspectives on supply chain risk management. *Production and Operations Management, 21*, 1–13.

Soyster, A. L. (1973). Convex programming with set-inclusive constraints and applications to inexact linear programming. *Operations Research, 21*, 1154–1157.

Sun, A., Davis, R., Starbuck, M., Ben-Amotz, A., Pate, R., & Pienkos, P. T. (2011). Comparative cost analysis of algal oil production for biofuels. *Energy*, *36*, 5169–5179.

Tang, C. S. (2006). Robust strategies for mitigating supply chain disruptions. *International Journal of Logistics: Research and Applications*, *9*, 33–45.

Tomlin, B. (2006). On the value of mitigation and contingency strategies for managing supply chain disruption risks. *Management Science*, *52*, 639–657.

Tu, Q., Eckelman, M., & Zimmerman, J. (2017). Meta-analysis and harmonization of life cycle assessment studies for algae biofuels. *Environmental Science & Technology*, *51*, 9419–9432.

Vilko, J. P., & Hallikas, J. M. (2012). Risk assessment in multimodal supply chains. *International Journal of Production Economics*, *140*, 586–595.

Wang, M., Huo, H., & Arora, S. (2011). Methods of dealing with co-products of biofuels in life-cycle analysis and consequent results within the US context. *Energy Policy*, *39*, 5726–5736.

Waters, D. (2011). *Supply chain risk management: Vulnerability and resilience in logistics*. Kogan Page Publishers.

Yatim, P., Lin, N. S., Lam, H. L., & Choy, E. A. (2017). Overview of the key risks in the pioneering stage of the Malaysian biomass industry. *Clean Technologies and Environmental Policy*, *19*, 1825–1839.

Yue, D., You, F., & Snyder, S. W. (2014). Biomass-to-bioenergy and biofuel supply chain optimization: Overview, key issues and challenges. *Computers & Chemical Engineering*, *66*, 36–56.

Ye, F., Hou, G., Li, Y., & Fu, S. (2018 a). Managing bioethanol supply chain resiliency: A risk-sharing model to mitigate yield uncertainty risk. *Industrial Management & Data Systems*, *118*, 1510–1527.

Ye, F., Li, Y., & Yang, Q. (2018 b). Designing coordination contract for biofuel supply chain in China. *Resources, Conservation and Recycling*, *128*, 306–314.

Yue, D., & You, F. (2016). Optimal supply chain design and operations under multi-scale uncertainties: Nested stochastic robust optimization modeling framework and solution algorithm. *AIChE Journal*, *62*, 3041–3055.

Zokaee, S., Jabbarzadeh, A., Fahimnia, B., & Sadjadi, S. J. (2017). Robust supply chain network design: An optimization model with real world application. *Annals of Operations Research*, *257*, 15–44.

CHAPTER 5

Sustainability concepts in biofuel supply chain

5.1 Supply chain management and introduction to sustainability paradigm

Different paradigms have been proposed in the supply chain management (SCM) literature in the last decades. Based on specific considerations and objectives incorporated in the supply chain optimal design and planning models, the SCM paradigms can be classified into lean, green, sustainable, agile/responsive, and resilient/robust, as depicted in Fig. 5.1. The lean supply chain paradigm is focused on removing any nonvalue adding activities, and therefore the main concern of a lean supply chain is economic considerations (Farahani, Rezapour, Drezner, & Fallah, 2014). Green supply chain management (GrSCM) seeks to integrate environmental thinking into the SCM to take into account environmental choices besides economic objective in the process of decision making (Srivastava 2008), excluding the social dimension of sustainability. A considerable part of GrSCM is related to reverse supply chains which include all processes and activities that are relevant to recycling, recovery, or disposal of the used products (Pishvaee, Razmi, & Torabi, 2014). As a more comprehensive paradigm over both lean and GrSCM, Ahi and Searcy (2015) describe the sustainable supply chain as an integration of economic, environmental, and social dimensions through interorganizational coordination to manage the three types of flows in supply chain, namely, material, information, and financial in an efficient and effective manner with the aim of improving the competitiveness, profitability, and resilience of the entire chain. Another well-konown paradigm of SCM is the responsive supply chain which is developed based on agile manufacturing concepts to respond to the current competitive marketplace in which the customers' needs and expectations are changing quickly. This agility can be achieved by reducing time to market through adopting new technologies and strategies in a cost-effective manner (Farahani et al., 2014).

In order to guarantee that the supply chain model is effective and practical for real-world applications, it must provide an adequate safeguard against uncertainty (Mohseni, Pishvaee, & Sahebi, 2016). To achieve this goal, as the

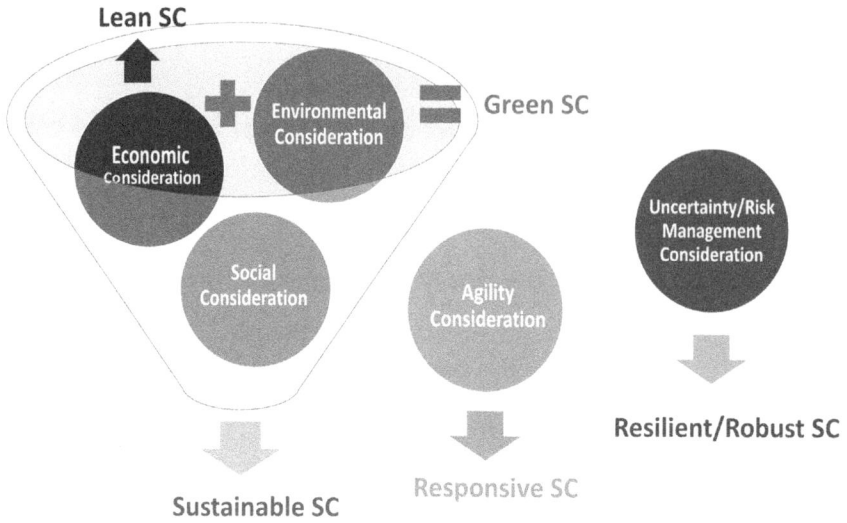

Figure 5.1 *Different types of objectives and considerations in supply chain design and planning models according to SCM paradigms.*

main concept of resilient/robust supply chain paradigm, there is a need to develop a comprehensive risk management framework that can be utilized for supply chain optimization models under uncertainty. Chapter 4 proposes a risk management framework for biomass supply chains that captures various dimensions of risk management to enable researchers to cope with a wide range of risks in three stages including risk identification, risk assessment, and risk treatment.

Most definitions of sustainable development agree that sustainability aims at ensuring two main elements including: (1) the existence of adequate supplies of the ecological, material, human, and social resources required to enable humans to meet their basic needs and also support continued development, and (2) equitable access to this adequate resources for members of the current generation (i.e., intergenerationally) as well as for the current and future generations (i.e., intragenerationally). Ecological resources consist of renewable sources (e.g., timber or food), and services (e.g., protection against ultraviolet solar radiation) that are provided by healthy natural ecosystems. Material resources include nonrenewable resources, while human resources refer to knowledge, income, human rights, health, freedom, and opportunity of applying the knowledge. Social resources address trust, equity, reciprocity norms, and other circumstances that facilitate the cooperation and coordination of mutual benefits (Gladwin 2001; Rodriguez, Roman, Sturhahn, & Terry, 2002).

One of the primary motivations of increasing attention to sustainable development is the environmental hardships caused by human activities. Nowadays, environmental issues are not merely regarded as a local or even national problem. However, with respect to the interdependency of environmental impacts with macro issues such as economy, culture, development, policy, and society, each environmental problem within the borders of the country is deemed as a global issue. In this regard, the immense environmental problems can be classified into global warming (greenhouse effect), ozone layer depletion, scarcity of freshwater resources, desertification and loss of fertile soil, air pollution and acid rain, toxic waste disposal, and the destruction of biodiversity (Hollander, 2003). On the other hand, a healthy human is recognized as a prerequisite of a healthy and dynamic society, and therefore, a vital element of sustainable development. However, a considerable number of people cannot afford basic sanitary and health services, which put human health and development processes at risk. Besides, poverty, unemployment, and inequality of opportunities to work according to geographic locations, gender, and age are other critical social concerns.

Accordingly, the increasing concerns about these environmental and social impacts of business activities have led to several international summits and finally, the development of the so-called "sustainable development" paradigm. Various definitions have been presented for sustainable development so far, however, the most widely accepted is that of the world commission on environment and development (WCED). In the published report by the commission, sustainable development is defined as "development that meets the needs of the present without compromising the ability of future generations to meet their own needs" (WCED, 1987). Based on this definition, sustainable development must set its objectives concurrently on three main dimensions including economic, environmental, and social, also referred to as three pillars of sustainability (Fig. 5.2), such that providing:

- Social growth to satisfy all individual's needs
- The effective protection of the environment and natural resources
- Sustainable economic growth

It is worth noting that incorporating all three aspects of sustainability into business is not limited to the boundary of a corporation. Instead, the sustainability should be ensured at the entire supply chain network. In addition, in order to ensure that biofuels can be used as a sustainable alternative to petroleum fuels, the economic, environmental, and social impacts throughout the biomass-to-biofuel supply chain must be incorporated in the design and planning of biomass supply chains. In this regard, supply

Figure 5.2 *The three dimensions of sustainability paradigm.*

chain sustainability has become an active area in the literature of biomass supply chains in recent years.

The remainder of this chapter is organized as follows. The economic aspect of sustainability and classification of economic objective functions applied in biomass supply chain optimization models as well as a brief description of financial ratios are presented in Section 5.2. Section 5.3 describes the life-cycle assessment (LCA) methodology for the sustainability analysis of biomass-to-biofuel supply chains. Moreover, the calculations of the LCA methodology steps for a biofuel supply chain according to Eco-indicator 99 method is demonstrated through a case study in this section. Section 5.4 presents the social impact assessment (SIA) as well as a brief review of SIA methods and guidelines. Finally, Section 5.5 is devoted to conclusions and future research directions.

5.2 Economic aspect

Due to the fact that the biofuel industry is still in the initial stages of its development (Yue, You, & Snyder, 2014a) and most of the biofuel production technologies are immature, the cost of biofuels (especially second- and third-generation biofuels) is significantly higher than the price of conventional fossil fuels. The economic aspect mainly focuses on developing a supply chain that manages the different activities of the chain from biomass production in fields to biofuel production in biorefineries and distribution to final markets in a cost–effective manner, aiming at improving the cost-competitiveness of biofuels. Therefore, reducing costs and redundancies as well as optimal utilization of resources are two main concerns of economic biofuel supply chains. This aspect has been addressed by a large number of studies in biofuel supply chain design and planning models by defining economic objective functions minimizing total costs or maximizing the total profit of the entire supply chain. The most common economic objective functions adopted can be classified into:

- Minimizing (expected) total costs/annual cost/unit cost (Gebreslassie, Yao, & You, 2012; Leduc, S., Natarajan, K., Dotzauer, E., McCallum, I., & Obersteiner, M., 2009; Li & Hu, 2014; Sharma, Ingalls, Jones, Huhnke, & Khanchi, 2013; Tong, Gleeson, Rong, & You, 2014),
- Maximizing (expected) total profit/annual income/annual profit/ net cash (Azadeh, A., Arani, H. V., & Dashti, H., 2014; Gonela, Zhang, Osmani, & Onyeaghala, 2015; Marufuzzaman, M., Ekşioğlu, S. D., & Hernandez, R., 2014; Osmani, A. & Zhang, J., 2013; Santibañez-Aguilar, Morales-Rodriguez, González-Campos, & Ponce-Ortega, 2016),
- Maximizing (expected) net present value (Giarola, Bezzo, & Shah, 2013; Kostin, Guillén-Gosálbez, Mele, Bagajewicz, & Jiménez, 2012; Marvin, W. A., Schmidt, L. D., Benjaafar, S., Tiffany, D. G., & Daoutidis, P., 2012; Osmani, A. & Zhang, J., 2014; Walther, Schatka, & Spengler, 2012),
- Minimizing investment/transportation/purchasing/operational costs (Aksoy et al., 2011; Balaman & Selim, 2015).

In addition to the above general economic metrics, there are financial ratios that can be employed as economic objective functions in biofuel supply chain models to optimize the efficiency of the supply chain in the utilization of its assets, which enhances the financial attractiveness of biofuel production and encourages investment in the biofuel industry. However, despite the emphasis on the importance of financial considerations in the SCM area, which is mentioned by many researchers (Longinidis & Georgiadis, 2011), there is a notable paucity of biomass supply chain design and planning models incorporating financial aspects. Financial ratios are broadly classified into the following categories (Ross, Westerfield, & Jordan, 2010):

- Short-term solvency or liquidity ratios: Ratios of this category represent the cash ability of the supply chain to pay its bill in the short-term.

These ratios concentrate on current assets and liabilities. The current ratio, cash ratio, and quick ratio are among the liquidity ratios.

- Long-term solvency or financial leverage ratios: This category addresses the long-term ability of the supply chain to fulfill its commitments in the future. Total debt ratio, cash coverage ratio, and times interest earned ratio are the measures of this category.
- Asset management or turnover ratios: The measures of this category, which are called asset utilization or turnover ratios, describe the efficiency and intensity of asset utilization to create sales. Inventory turnover and day's sales in inventory are two prominent examples of turnover ratios.
- Profitability ratios: Profitability measures, which are the most important and most widely employed financial ratios, represent the efficiency of the supply chain in the utilization of its assets and managing its operations. Profit margin, return on assets, and return on equity are among profitability ratios. It should be noted that the focus of this category is on net income.
- Market value ratios: The information of this group of measures is not necessary for financial statements. Definitely, the calculations of these measures can be performed merely for publicly traded firms. Price-Earnings ratio and market-to-book ratio are the measures of market value.

5.3 Environmental impacts assessment: LCA methodology

LCA is the most used methodology to quantify and evaluate the environmental impacts associated with a product or service across its life cycle stages (i.e., cradle to grave scope) such as the acquisition of raw materials, production, distribution, use, recycling and disposal. The most credible framework for LCA has been proposed by the International Organization for Standardization (ISO), which has four main phases, namely, goal and scope definition, inventory analysis, impact assessment and interpretation. Fig. 5.3 depicts LCA framework phases based on ISO 14040:2006 and 14044:2006 standard series (ISO, 2006). Descriptions of LCA phases are given in the following sections.

5.3.1 Goal and scope definition phase

In this phase, the purpose of applying the LCA study, a description of the system boundaries and application, and a definition of the functional unit

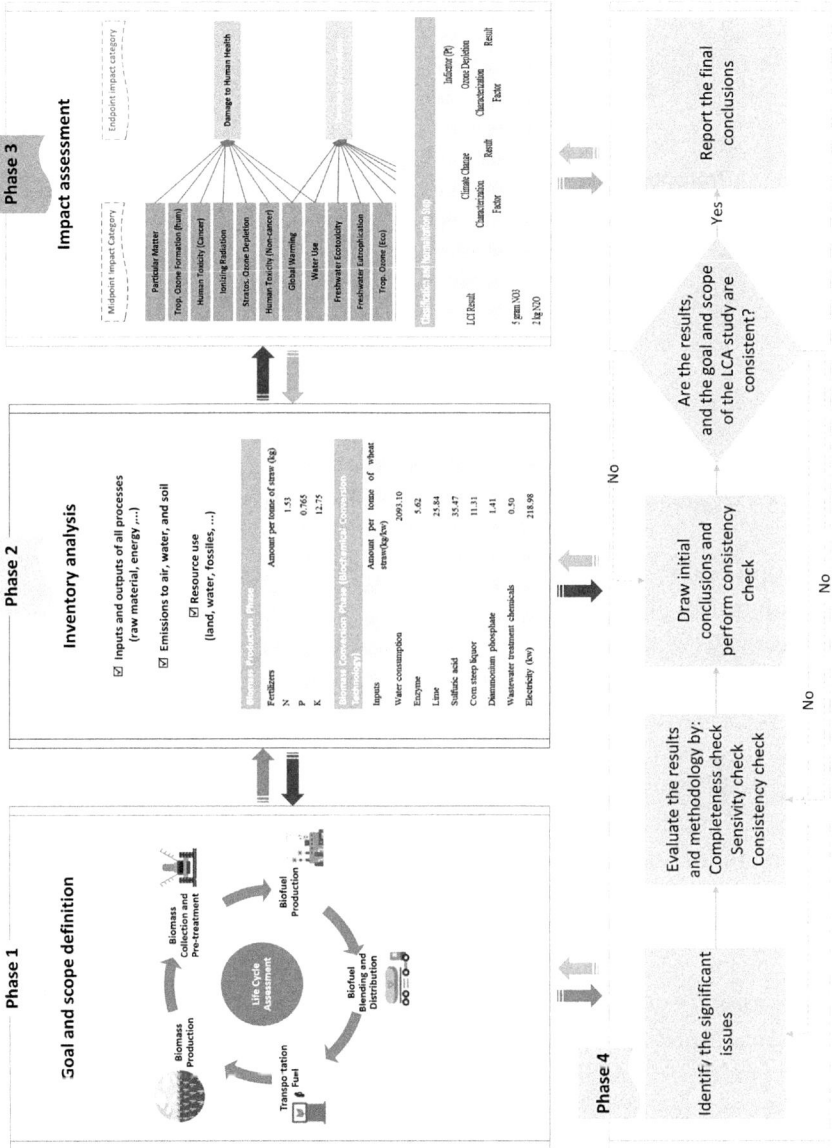

Figure 5.3 *Conceptual framework of LCA study according to ISO 14040/44 standard series.*

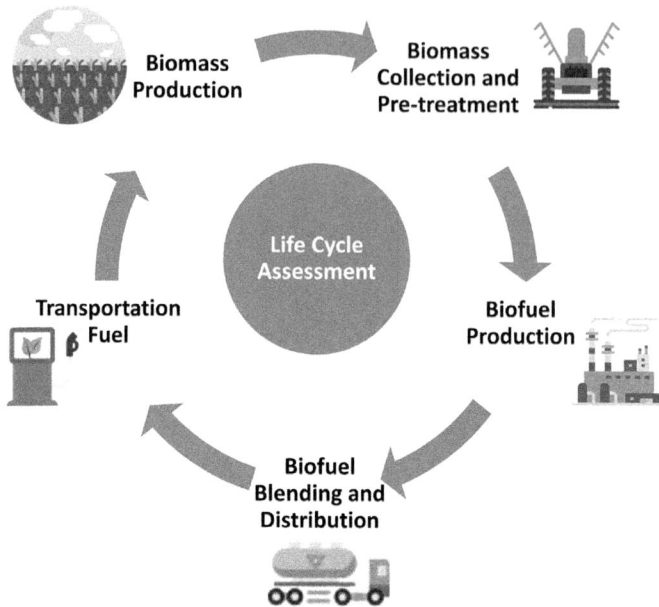

Figure 5.4 *Cradle to grave system boundary in biofuel life cycle assessment.*

a reference unit for all life cycle calculations are provided. According to the boundaries of a supply chain, four scopes for LCA study can be identified, namely, "cradle-to-grave", "cradle-to-gate", "gate-to-gate," and "gate-to-grave". The system boundaries in the "cradle-to-grave" scope concern all stages across the supply chain from raw materials acquisition to final disposal and recycling. The cradle-to-grave LCA study, which is called field-to-wheel in the context of biomass-to-biofuel supply chains, consists of biomass cultivation, biomass harvesting and handling, biomass pretreatment, biofuel production, biomass and biofuel storage and transportation, and fuel use, as depicted in Fig. 5.4. The cradle-to-gate boundary addresses the life cycle stages from raw material extraction to the factory gate. The cradle-to-gate scope is called "field-to-tank" in the biomass-to-biofuel supply chains, which involves supply chain stages from biomass cultivation to biofuel production in biorefineries until delivery to intermediate facilities for blending with counterpart fossil fuels and/or distribution to end-users.

The gate-to-gate approach focuses only on production processes and excludes the upstream and downstream steps related to raw material extraction and disposal. This boundary is further adopted in recovery networks that transform end-of-life products to be re-used in another or the same supply chain. Finally, the gate-to-grave scope is limited to the ultimate steps of a supply chain, ranging from factory gate to product disposal, which is handled by waste or reverse supply chains (Eskandarpour, Dejax, Miemczyk, & Omega, 2015).

As mentioned above, in addition to the scope definition, a functional unit should also be defined in the first phase of an LCA study. A functional unit is a standard unit of all calculations which are performed in different steps of the LCA study. For example, for measuring and analyzing the environmental impacts of biomass production in biomass to biofuel supply chains, 1 ton of biomass can be determined as a functional unit.

5.3.2 Life cycle inventory analysis phase

In the second phase of the LCA study, the inventories associated with each life cycle stage should be complied and quantified with regard to the defined system boundary. Life cycle inventory (LCI) methodology determines input and output flows of a product across its whole life cycle stages. Such flows involve raw material, energy, and other physical resources from the environment as inputs, as well as emissions to air, soil, and water as outputs. The LCI data of many processes required to implement the inventory analysis phase are available in various well-known life cycle inventory databases, such as ecoinvent (ecoinvent, 2019), Greet (Greet, 2019), GaBi (GaBi, 2019), U.S. LCI (NREL, 2012), and openLCA (openLCA, 2012).

5.3.3 Life cycle impact assessment phase

In the life cycle impact assessment (LCIA) phase, LCI results are analyzed according to their potential contributions to specified impact categories, which are taken into account in the LCA study. According to ISO14040 and ISO 14044 (ISO, 2006), LCIA phase is composed of three mandatory steps, namely selection, classification, and characterization, and two optional steps, namely normalization, and weighting.

- **Selection of impact categories, indicators and characterization models (selection)**

    In the first step of this phase, impact categories and associated indicators are selected based on the applied LCIA method. Definition and examples of common terms used in ISO 14040/14044 are depicted in Table 5.1.

    It should be noted that according to the employed impact categories, LCIA methods are divided into midpoint-oriented and endpoint-oriented methods. Midpoint-oriented methods address environmental impacts in various impact categories such as climate change, terrestrial acidification, ozone depletion, human toxicity, and agricultural land occupation, while

Table 5.1 Definition and examples of common terms in ISO 14040/14044 (ISO, 2006).

Term	Definition	Examples		
(midpoint) Impact category	The environmental concerns to which inventory results may contribute to.	Climate change	Ozone depletion	Acidification
Category Indicators	Quantitative indication of the related impact category	Infrared radiative forcing increase	Stratospheric ozone decrease	
Characterization factor (CF)	A factor used to convert inventory results to the unit of the related category indicator	Global warming potential (GWP)	Ozone depletion potential (ODP)	Acidification potential (AP)
Unit	Common unit of impact category	kg CO_2-eq to air	kg CFC-11-eq to air	kg SO_2-equivalents

endpoint-oriented methods aggregate the midpoint level categories in limited damage categories, such as damage to ecosystem quality and damage to human health (Huijbregts et al., 2017).

- **Assignment of LCI results to the impact categories (classification) and calculation of category indicator results (characterization)**

  In the classification step, the inventory results such as resource consumption, and emissions into air or water are assigned to the impact categories according to the environmental problems in the respective category. In the characterization step, LCI results are multiplied with a characterization factor and summed to obtain the impact score of each impact category in terms of its common units. Fig. 5.5 illustrates calculations of classification and characterization steps for three impact categories based on the ReCiPe2016 (the Hierarchist version) (Huijbregts et al., 2017). Note that the results of impact categories are not comparable because each category has its own unit.

Classification Step

LCI Result	Impact Categories		
	Climate Change	Ozone Depletion	Acidification
2 kg CO_2 (Carbon dioxide)	✓		
10 gram N_2O (Nitrous oxide)	✓	✓	
5 gram NH_3 (Ammonia)			✓

Classification and Normalization Step

LCI Result	Indicator (Pt)					
	Climate Change		Ozone Depletion		Acidification	
	Characterization Factor	Result	Characterization Factor	Result	Characterization Factor	Result
2 kg CO_2 (Carbon dioxide)	1	2				
10 gram N_2O (Nitrous oxide)	298	2.98	0.011	1.10E-04		
5 gram NH_3 (Ammonia)					1.96	9.80E-03
Impact Category	Climate Change		Ozone Depletion		Acidification	
Indicator Result	4.98 kg CO2-eq to air		1.10E-04 kg CFC-11-eq to air		9.80E-03 kg SO2-equivalents	
Normalization Value	1.12E+4 kg CO2/yr		2.20E-2 kg CFC11/yr		4.15E-1 kg P/yr	
Normalized Value (yr)	4.45E-04		5.00E-03		2.36E-02	

(10/1000)×298

4.98/1.12E+4

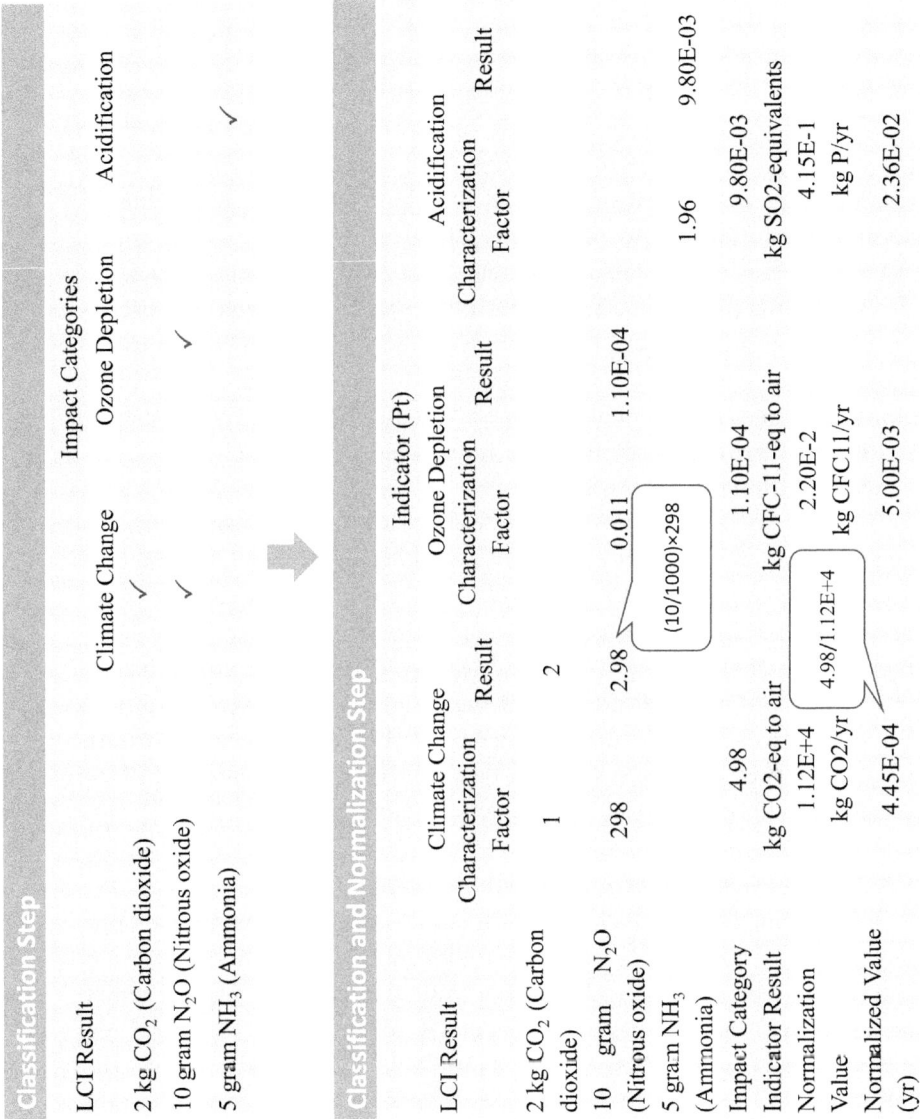

Figure 5.5 *Classification, characterization, and normalization steps in the LCIA phase of LCA study.*

- **Normalization and weighting**

Normalization and weighting steps are optional in ISO 14040/44 standard series. In order to specify the general magnitude of the score of each impact category indicator obtained in the characterization step and compare the scores of different impact categories with different metrics, a normalization step can be adopted in the LCA study. Normalization means that the score of each impact category indicator is divided by the specified reference values. Moreover, the normalized scores can be used for weighting the results of the category indicators. It should be noted that different normalization sets can be adopted as reference values. A common scale for the normalization reference set is the average annual impact of a person for each impact category. As depicted in Fig. 5.5, the average annual impact of a European in the year 2000 is selected as reference normalization values, that is, an average European citizen emits approximately 11 tons of CO_2, 22 g of CFC11, and 0.415 kg of phosphate equivalent to water (Goedkoop et al., 2016). Accordingly, normalized results are obtained by dividing the indicator results of each impact category into the reference values.

In order to facilitate the ultimate comparison of indicator results obtained for different impact categories and interpretation of LCIA method results, the normalized indicator results of different impact categories can be weighted according to specified weighting factors. Indeed, the weighting factors assigned to each impact category represent the importance of that category. It is worth noting that the weighting is a subjective process because the weighting factors are selected based on decision maker's priorities. Finally, the weighted results can be aggregated, and therefore, the final result of the LCA study can be provided as a single score.

5.3.3.1 Life cycle impact assessment methods

In this section, the list and characteristics of various LCIA methods are presented. As mentioned before, LCIA methods are classified into the midpoint and endpoint methods according to the covered impact categories. It should be noted that some of these methods utilize normalization and weighting methods and provide a single final score. Table 5.2 illustrates the characteristics of the various LCIA methods according to the covered impact categories and the option of providing a final score. Among the listed midpoint methods, CML2001 and TRACI, and among the endpoint methods, Eco-indicator 99 and both versions of ReCiPe, which are the most credible methods of their category, are described in the subsequent paragraphs.

Table 5.2 Characteristics of credible LCIA methods.

LCIA method	Author	Covered impact categories		
		Midpoint-oriented	Endpoint-oriented	Providing a final score
Eco-indicator 99	(Goedkoop, M. & Spriensma, R., 2000)	√	√	√
CML2001	Guinée (2001)	√	–	–
EDIP 2003	(Hauschild, M. & Potting, J., 2005)	√	–	–
EPS 2000	Steen, 1999	√	√	–
IMPACT 2002+	Jolliet et al., 2003	√	√	–
Ecological Scarcity	Brand, Braunschweig, Scheidegger, & Schwank, 1998	√	√	–
TRACI 2.0	Bare (2011)	√	–	–
ReCiPe 2008	Goedkoop et al. (2009)	√	√	√
ReCiPe 2016	Huijbregts et al. (2017)	√	√	√

- **Eco-indicator 99 (EI99)**

Eco-indicator 99 (EI99) is a damage-oriented LCIA method, which is definitely one of the most widely used LCIA methods. The first edition of EI99 (Goedkoop, Hofstetter, Müller-Wenk, & Spriemsma, 1998) is introduced by a group of scientists in 1997, which replaced Eco-indicator 95 methodology. Afterward, two editions of the final report have been presented, and finally the third edition of EI99 (Goedkoop, M. & Spriensma, R., 2000) has been made available in 2001. EI99 as an endpoint-oriented impacts assessment method quantifies and analyzes environmental impacts in three damage categories, including (1) damage to human health, (2) damage to ecosystem quality, and (3) damage to resources. It should be noted that in this method, midpoint and endpoint impact categories are associated, and indeed, EI99 covers both category levels. To quantify the impacts, EI99 employs special units in all categories, namely eco-indicators. More specifically, eco-indicator point (Pt) is the standard unit of measurement in this method, which is broken down into 1000 millipoints (mPt).

One of the most practical advantages of this method is that EI99 presents a weighting method that allows the expression of LCA results as a single final score that can be used to compare all the studied alternatives based on their environmental impacts. Assuming different weighting percentages for each damage category, the damages (eco-indicators) are weighted according to three cultural perspectives, including (1) hierarchist (default), (2) individualist, and (3) egalitarian. For instance, the hierarchist perspective assumes 20%, 40%, and 40% for resource depletion, human health, and ecosystem quality damage categories, respectively. The extension of this method for simplification of the interpretation phase and weighting of results led to the emergence of other LCIA methods such as IMPACT 2002+ and ReCiPe2008.

- **CML2001**

CML2001, as a midpoint-oriented impact assessment method, has been published by the center of Environmental Science of Leiden University (CML). This method presents an operational guide for implementing the four-phase LCA framework, which is proposed by the ISO 14040 series. In this method, emission categories are classified into baseline (i.e., the most common factors such as climate change), and study-specific (such as freshwater sediment ecotoxicity) impact categories. One shortcoming of CML2001 is that although the relation between midpoint and endpoint indicators has been specified, a quantification method for evaluating endpoint impacts is not presented. On the other hand, despite that some possible normalization methods to quantify the results have been introduced in CML2001, no clear method for weighting mechanism and calculating a final score is provided.

- **ReCiPe2008**

ReCiPe2008 is one of the most comprehensive LCIA methods that is able to assess environmental impacts at both midpoint and endpoint categories. This damage-oriented method first calculates life cycle inventories and their environmental impacts according to 18 midpoint indicators. The midpoint impacts are then aggregated in three endpoint indicators, including damage to (1) human health, (2) ecosystems, and (3) resource availability. Eventually, by utilizing a weighting method, the final result of total environmental impacts is achieved and presented as a single final score. Three distinctive cultural perspectives are applied in the ReCiPe for weighting environmental impacts results and calculating a final score, including hierarchist, the individualist, and egalitarian, which are exactly identical to those of the EI99 method.

It should be noted that the ReCiPe method is developed based on the CML2001 and EI99 methods, and therefore covers the advantages and strengths of both methods. In order to assess midpoint and endpoint environmental impacts, the ReCiPe method adopts experiences related to the CML2001 and EI99, respectively (Goedkoop et al., 2009; Pishvaee et al., 2014).

- **TRACI**

In order to conduct various LCA case studies, the U.S. Environmental Protection Agency's National Risk Management Research Laboratory presented a review of existing LCA methodologies such as Eco-indicator 99, CML2001, and IMPACT 2002+, and finally decided to develop an impact assessment software by applying the best methodologies in each impact category. Accordingly, the original version of TRACI (the Tool for the Reduction and Assessment of Chemical and other Environmental Impacts) methodology (Bare, Norris, Pennington, & McKone, 2002) was released in August 2002 (Bare, 2012).

TRACI 2.0, as an updated version of the original version of TRACI, is a midpoint oriented LCIA methodology which has been released in 2011. This method investigates environmental impacts in ten midpoint impact categories, as depicted in Table 5.3. It should be noted that ongoing research has focused on quantifying the land use and water use impact indicators in future versions of TRACI. Moreover, the considered impact categories have not been aggregated in TRACI, and therefore, no normalization or weighting methods are presented in this method (Bare, 2011, 2012).

- **ReCiPe2016**

ReCiPe2016 method is an updated version of the ReCiPe2008 method (Goedkoop et al., 2009), which has been developed by Huijbregts et al., (2017). This method incorporates three areas of protection, namely, human health, resource scarcity, and ecosystem quality (including terrestrial, freshwater, and marine ecosystems) as endpoint impact categories, while

Table 5.3 Midpoint impact categories of TRACI 2.0 (Bare, 2011, Bare, 2012).

Impact categories	
1. Global warming	2. Photochemical smog formation
3. Acidification	4. Human health particulate effects
5. Eutrophication	6. Human health cancer
7. Ecotoxicity	8. Human health noncancer
9. Ozone depletion	10. Fossil fuel depletion effects

including 17 midpoint impact categories. In comparison with the ReCi-Pe2008, the impact categories in ReCiPe2016 make reference to the global scale rather than the European scale. However, the possibility of implementing characterization factors at country and continental scales is provided in some impact categories. Furthermore, the ReCiPe2016 method proposes new damage pathways, including water use impacts on human health, water use and tropospheric ozone formation on terrestrial ecosystems, and water use and climate change impacts on freshwater ecosystems. It is noteworthy that although no single score weighting method is proposed in this method, the weighting method of the previous version can be applied (Huijbregts et al., 2017).

5.3.3.2 Case study

In order to conduct an LCA study for a case study of Bairamzadeh et al., (2015), the calculations of the LCA steps for a bioethanol supply chain according to Eco-indicator 99 method are presented in this section. The system boundary of the concerned bioethanol supply chain is depicted in Fig. 5.6, which is based on a cradle to grave scope. According to Fig. 5.6, the life cycle phases of the supply chain include feedstock production,

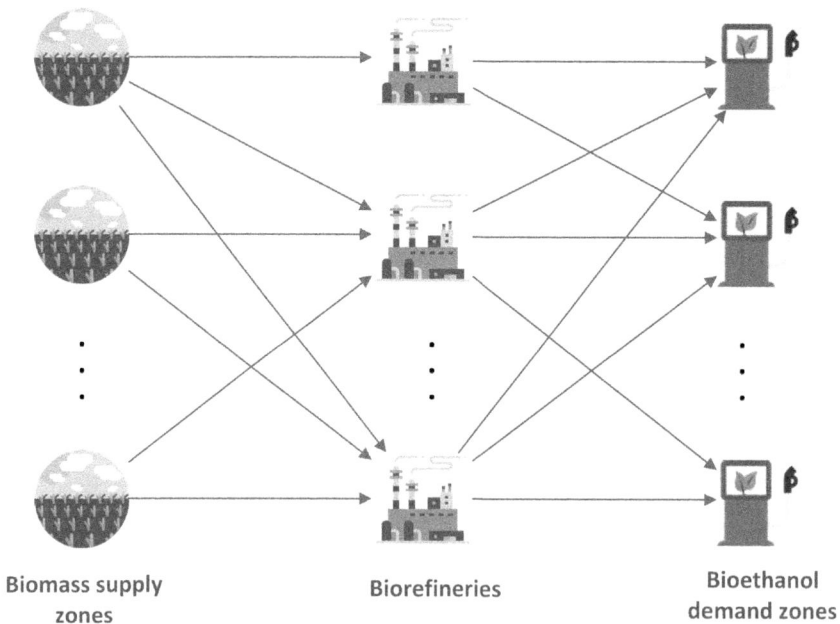

| Biomass supply zones | Biorefineries | Bioethanol demand zones |

Figure 5.6 *System boundary of the concerned bioethanol supply chain.*

converting biomass into bioethanol, transportation of biomass and bioethanol, and bioethanol use phase. It should be noted that based on Mu, Seager, Rao, & Zhao, (2010) fuel use phase is excluded from environmental impacts calculations because the carbon dioxide, which is released during the utilization of bioethanol, does not have eutrophication and acidification impacts. Besides, in order to divide the impacts among the product and the coproducts in the environmental assessment of a system that produces more than one output, an allocation method must be adopted. Accordingly, the subdivision method is applied to calculate environmental impacts in the biomass production phase, which considers only additional activities caused by the wheat straw harvest in this phase. The list of inventory results and calculations of environmental impacts in each phase are summarized in Figs. 5.7 and 5.8.

Biomass production phase	
fertilizers	Amount per ton of wheat straw (kg)
N	1.53
P	0.765
K	12.75

Biomass conversion phase (Biochemical conversion technology)	
Inputs	Amount per ton of wheat straw (kg/kw)
Water consumption	2093.10
Enzyme	5.62
Lime	25.84
Sulfuric acid	35.47
Corn steep liquor	11.31
Diammonium phosphate	1.41
Wastewater treatment chemicals	0.50
Electricity (kw)	218.98
Solid wastes	73.44
Carbon dioxide	1040.53

Figure 5.7 *Life cycle inventory analysis (LCI) phase of the concerned bioethanol supply chain.*

Biomass Production Phase

LCI Result (Required fertilizers)	Amount per Functional Unit (1 ton wheat straw)	Indicator (Pt)			Result
		Human health	Ecosystem Quality	Resources	
N2O (kg)	0.0686 ✖	1.79	-	-	0.12
NO (kg)	0.0107	3.36	0.69	-	0.05
NO3 (kg)	0.2899	-	0.45	-	0.13
NH3 (kg)	0.0306	2.21	1.22	-	0.10
Total (Pt)					0.40

Biomass Conversion Phase (Biochemical Conversion Technology)

LCI Result (Required inputs)	Amount per Functional Unit (1 ton wheat straw)	Indicator (Pt)			Result
		Human health	Ecosystem Quality	Resources	
Sulfuric acid (kg)	35.47	0.0374			1.33
Diammonium phosphate (kg)	1.4132	0.272			0.38
Urea (kg)	209.31	0.318			66.56
NAOH (kg)	1883.79	0.0517			97.39
Lime (kg)	25.84	0.0199			0.51
Heat (MJ)	76.88	0.00494			0.38
Total (Pt)					166.56

Biomass and Biofuel Transportation Phase

Amount per Functional Unit (1 ton wheat straw/1 gallon bioethanol)	Indicator (Pt)			Result
	Human health	Ecosystem Quality	Resources	
Truck 28 ton	0.0219			0.0219
Boiler Truck 8000 gallon	0.000114			0.000114

Figure 5.8 *Cradle to grave LCA study calculations of the concerned bioethanol supply chain.*

5.3.4 Life cycle interpretation phase

The last phase of an LCA study involves the interpretation and analysis of the results obtained from the previous phases to provide a set of conclusions and recommendations according to the defined objective and scope. Based on ISO 14044:2006, the main steps of this phase are summarized as follows (ISO 14044 2006):

1. Identify the significant issues relying on the results of the life cycle inventory analysis and impact assessment phases of LCA. Examples of these issues can be related to inventory data (e.g., emissions, energy, and waste), impact categories (e.g., climate change), and contributions from life cycle stages, such as groups of processes or individual unit process (e.g., energy production or transportation).

2. Evaluate the results and methodology in terms of completeness, sensitivity, and consistency. An evaluation of LCA study performance is conducted for ensuring the reliability of results. The following techniques can be used for the performance evaluation:

   a. Completeness check: The purpose of this check is to guarantee the availability and completeness of all related information for interpretation.

   b. Sensitivity check: The purpose of this analysis is to assess the reliability of the LCA results regarding the different types of uncertainties due to the uncertain input data, different allocation rules, considered impact categories, etc.

   c. Consistency check: The purpose of this check is to ensure the consistency of the data, methods, and assumptions with the objective and scope of the LCA study considering the requirements of data quality, predefined values and assumptions, methodological limitations, etc.

3. Draw initial conclusions and perform the consistency check to evaluate the conclusions

4. If the consistency check is verified, report the conclusions, otherwise, go back to one of the appropriate previous steps. It should be noted that the purpose of this step is to draw conclusions, determine limitations, and provide recommendations.

The main procedure of the interpretation phase and interactions of this phase with other phases in the LCA framework, based on ISO 140044 standard, are depicted in Fig. 5.3.

5.4 Social impact assessment

The rapidly growing attention to the social impacts in firms and corporation's decisions has led to emerging the concept of corporate social responsibility (CSR) (Ciliberti, Pontrandolfo, & Scozzi, 2008). World business council for sustainable development (WBCSD) describes social responsibility as the "continuing commitment by business to behave ethically and contribute to economic development while improving the quality of life of the workforce and their families as well as of the local community and society at large" (Holme, R. & Watts, P., 1999). In the same way, the Commission of European Communities (CSR & RBC, 2019) describes CSR as "the responsibility of enterprises for their impact on society" which can be achieved by companies through "integrating social, environmental, ethical, consumer, and human rights concerns into their business strategy and operations" as well as "following the law". Accordingly, in the literature of CSR, environmental issues are considered as a subset of social responsibility. However, in the sustainability paradigm, the environmental dimension is of great importance and is assumed to be one pillar of sustainable development.

Despite the extensive literature on the environmental aspect of sustainability in the supply chain design and planning problems, few studies have incorporated the social aspect of the sustainability paradigm (Bairamzadeh, Pishvaee, & Saidi-Mehrabad, 2015). Actually, the sustainability paradigm still devotes more attention to the environmental aspect in comparison with social responsibility (SR) that takes into account environmental issues as a subset of social concerns (Pishvaee et al., 2014). Accordingly, the World Summit on Sustainable Development (WSSD) held in 2002 re-emphasized that the sustainable development involves the balancing of not only economic and environmental aspects, but also social dimension, which are called three pillars of sustainability (White and Lee, 2009). The social dimension, as the third pillar of sustainability, incorporates measuring and analyzing all aspects of social impacts (SIs) through the entire supply chain. Several standards have been developed to present a framework for the implementation of SR, such as ISO 26000, SA8000, and AA1000. As one of the most adopted frameworks for social assessment, ISO 26000 categorizes the core subjects of SR into seven groups, each of which involves several issues concerning SR, as illustrated in Table 5.4.

5.4.1 Social impact assessment in biomass supply chains

There is a growing body of literature that recognizes the importance of the social performance of biofuel supply chains. Domac, Richards, and Risovic (2005) categorized the socio-economic aspects of bioenergy sector into four main dimensions including social benefits associated with each dimension, as follows: (1) social aspects dimension which leads to the social cohesion and stability and increased standard of living, (2) macrolevel dimension with benefits such as regional growth and security of supply/risk diversification, (3) supply-side dimension with benefits such as enhanced competitiveness and increased productivity, and (4) demand-side dimension which leads to benefits such as employment and income and wealth creation. The findings of their study showed that bioenergy, as one of the most labor intensive industry, has the greatest potential of job creation among other renewables in terms of local, regional as well as national employment. In a review paper, Awudu and Zhang (2012) classified the social issues of biofuel development into three groups including (1) potential of poverty reduction, (2) impacts on land and crop, and (3) impacts on social resources such as water utility systems. In this regard, biofuel is expected to contribute to poverty reduction by providing energy, improving the income and economic per capita, and enhancing the quality of life in the related communities. The second social

Table 5.4 ISO 26000:2010 social responsibility categories and associated impact subcategories (ISO, 2010b).

Social responsibility categories	Impact subcategories
Organizational governance	–
Human rights	Due diligence
	Human rights risk situations
	Avoidance of complicity
	Resolving grievances
	Discrimination and vulnerable groups
	Civil and political rights
	Economic, social and cultural rights
	Fundamental principles and rights at work
Labor practices	Employment and employment relationships
	Conditions of work and social protection
	Social dialogue
	Health and safety at work
	Human development and training in the workplace
The environment	Prevention of pollution
	Sustainable resource use
	Climate change mitigation and adaptation
	Protection of the environment, biodiversity and restoration of natural habitats
Fair operating practices	Anticorruption
	Responsible political involvement
	Fair competition
	Promoting social responsibility in the value chain
	Respect for property rights
Consumer issues	Fair marketing
	factual and unbiased information and fair contractual practices
	Protecting consumers' health and safety
	Sustainable consumption
	Consumer service, support, and complaint and dispute resolution
	Consumer data protection and privacy
	Access to essential services
	Education and awareness
Community involvement and development	Community involvement
	Education and culture
	Employment creation and skills development
	Technology development and access
	Wealth and income creation
	Health
	Social investment

issue arising from the fact that vast tracts of land are required for biomass production. Therefore, biofuel development can have an indirect impact on the land ownership system as well as the selection of appropriate feedstocks for cultivation. Moreover, the farmer's decisions about total land devoted to biomass feedstocks cultivation can affect the price and demand for agricultural products. Finally, the third social issue points out that the rapid increase of large scale biomass feedstock plantations for biofuel production leads to pollution and inadequacy of water in the short term. Recently, Dale, Kline, Richard, Karlen, & Belden, (2018) determined 16 socio–economic indicators relevant to cellulosic biofuels, which are classified into six socio-economic sustainability categories, namely, energy security, social wellbeing, resource conservation, external trade, profitability, and social acceptability. Despite the fact that the development of the biofuel industry has numerous social implications, only a few biofuel supply chain design and planning studies with a limited number of social criteria have been conducted. On the other hand, a review of the literature shows that maximizing the number of jobs created through the biomass to biofuel value chain is the most common social objective function in previous studies that address the social performance of biomass supply chains, such as (Bairamzadeh et al., (2015); Chávez, Sarache, & Costa, (2018); Santibañez-Aguilar, González-Campos, Ponce-Ortega, Serna-González, & El-Halwagi, (2014); You, Tao, Graziano, & Snyder, (2012); Yue, Slivinsky, Sumpter, & You, (2014b)). This research gap highlights the need for considering and optimizing social impacts associated with different phases of biofuel development from construction to operation of biomass supply chains in future studies.

5.4.2 Social life cycle assessment methods and guidelines

A brief review of social life cycle assessment methods and guidelines that are developed and published by researchers for the measurement and implementation of SR is provided in this section. Notably, unlike the environmental impact assessment methods, which present a practically precise computational and analytical procedure, social life cycle assessment methods are weak in this area.

- **Social accountability 8000 (SA8000) standard**

SA8000 standard is a social certification standard published for organizations and factories by social accountability international (SAI) in 1997. SA8000:2014 is the last (fourth) version of this standard, which replaced SA8000:2008 version in June 2014. This standard mainly focuses on social accountability issues, and specifically labor provisions in workplaces, which

Table 5.5 Social accountability indicators of SA8000 standard (SAI, 2014).

Social performance indicators	
1. Child Labor	2. Forced or Compulsory Labor
3. Working Hours	4. Health and Safety
5. Freedom of Association and Right to Collective Bargaining	6. Disciplinary Practices
7. Discrimination	8. Remuneration
9. Management System	

is developed based on the International Labor Organization (ILO) and the Universal Declaration of Human Rights conventions. Accordingly, SA8000 measures the social performance of organizations and factories in nine areas, as illustrated in Table 5.5 (SAI, 2014).

- **Global Reporting Initiative Sustainability Reporting Standards**
  Global Reporting Initiative (GRI) was founded by the coalition for environmentally responsible economies (CERES) and the Tellus Institute with the accompaniment of the United Nations environment program (UNEP) in Boston, USA in 1997. GRI presents a set of interrelated and modular standards, namely, GRI Sustainability Reporting Standards (GRI standards) which can be utilized by organizations to prepare a sustainability report. The sustainability report contains an organization's economic, environmental, and social impacts and therefore demonstrates its contributions to sustainable development. The GRI standards are presented as universal and topic-specific standards that are classified into four series including 100 series, GRI200, GRI300, and GRI400. The overview of the current version of GRI standards published in 2016 is given in the following (GRI, 2016):
- 100 series (Universal standards)
  - GRI101 (foundation): GRI101 specifies the reporting principles to ensure that the sustainability report is in accordance with the GRI standards in terms of report content and quality. Report content principles determine the required content to be included in the report considering activities, impacts, and interests of stakeholders. Report quality principles ensure the quality of information given in a sustainability report in terms of accuracy, clarity, and reliability.
  - GRI102 (general disclosures): GRI102 includes the required contextual information about the organization, such as its strategy, profile, ethics, and stakeholder engagement activities.
  - GRI103 (management approach): GRI103 reports the management approach for economic, environmental, and social topics in a

sustainability report such as the boundary of the topic and the corresponding impacts.

- GRI200 (economic topics): GRI200 standard series report specific disclosures about economic topics of an organization which are presented in GRI201–GRI206 standard parts. These topics include economic performance (GRI201), market presence (GRI202), indirect economic impacts (GRI203), procurement practices (GRI204), anticorruption (GRI205), and anticompetitive behavior (GRI206). For instance, GRI 201 addresses the direct economic value generated and distributed, defined benefit plan obligations, financial implications of climate change, and government financial assistance.

- GRI300 (environmental topics): GRI300 standard series report specific disclosures about environmental topics of an organization which are presented in GRI301–GRI308 standard parts. These topics include materials (GRI301), energy (GRI302), water-and-effluents (GRI303), biodiversity (GRI304), emissions (GRI305), effluents-and-waste, environmental compliance (GRI306), supplier environmental assessment (GRI307). For instance, GRI 301 addresses recycled input materials used, materials used by weight or volume, reclaimed products, and their packaging materials.

- GRI400 (social topics): GRI400 standard series report specific disclosures about social topics of an organization which are presented in GRI401–GRI419 standard parts. These topics include but not limited to employment (GRI401), labor-management relations (GRI402), occupational-health-and-safety (GRI403), training and education (GRI404), and diversity-and-equal-opportunity (GRI405). For instance, GRI401 addresses new employee hires and employee turnover, parental leave, and benefits provided to full-time employees.

- **Ethical trading initiative (ETI)**

The Ethical trading initiative was founded in 1998 as a not-for-profit company by a number of British retailers, aid agencies, workers' rights organizations, and trade unions to improve fundamental human rights in global supply chains. The main goal of this initiative is to eliminate exploitation and discrimination, and provide freedom, security, and equity for all workers in the world. To this aim, this initiative has developed the ETI base code which consists of nine clauses such as "employment is freely chosen" that takes into account the most relevant conventions of the International Labor Organization about labor practices. Besides, ETI has developed principles of implementation (POI) framework as a guide for members in order to implement the base code in their supply chains. These principles are classified into four areas including "commitment", "identifying labor rights issues", "prevent, mitigate, remedy", and "track & communicate". It should be noted that no measures are provided for the base code or principles of implantation by ETI (ETI, 2018).

- **Fair labor association (FLA)**

  FLA was founded in 1999 by a collaborative effort of colleges and universities, civil society organizations, and socially responsible companies. The main goal of FLA as an international organization is to create solutions to prevent abusive labor practices through providing training, tools and resources, and due diligence by means of independent assessments (FLA, 2019). FLA has developed a workplace code of contact as labor standards based on the International Labor Organization (ILO) standards and also internationally accepted labor practices. It should be noted that FLA monitors the compliance of affiliated companies with the principles of the workplace code to check the adherence to the principles of monitoring and compliance benchmarks. The compliance benchmarks specify the requirements of satisfying each standard, where the principles of monitoring address the procedure of compliance assessment. The workplace code of conduct presents nine elements including "employment", "forced labor", "child labor", "relationship", "nondiscrimination", "harassment or abuse", "freedom of association and collective bargaining", "health, safety, and environment", "hours of work", and "compensation" (WCC, 2019).

- **United Nations Global Compact (UNGC)**

  Global Compact (GC) is a voluntary initiative of the UN which is launched in 2000 to implement universal sustainability principles aiming at supporting the UN's goals. The UN Global Compact has developed a framework for corporate sustainability including ten principles in four main areas of human rights, environment, labor, and anti-corruption. For instance, the human rights area involves two principles as follows:

- Human rights:
  - "Principle 1: businesses should support and respect the protection of internationally proclaimed human rights", and
  - "Principle 2: make sure that they are not complicit in human rights abuses."

It is worth noting that the ten principles of the UN Global Compact are derived from international standards and labor practices such as the Universal Declaration of Human Rights and the International Labor Organization's Declaration on Fundamental Principles and Rights at Work

(UNGC, 2019). Notably, despite the presented sustainability principles and successful benchmarks, the GC framework does not refer to any measures for sustainability assessment.

- **Guidelines for Social Life Cycle Assessment of Products (GSLCAP)**

  GSLCAP have been published by the United Nations Environment Program (UNEP) in 2009. This guideline, as the only product-oriented (vs. organization oriented) SIA framework, presents a method for the assessment of social and socio-economic impacts of products through their life cycle stages. Theoretically, GSLCAP is developed based on the concepts of Social LCA (S-LCA) and stakeholder theory. S-LCA assesses the social and socio-economic impacts associated with a product across its life cycle stages (i.e., cradle to grave scope). Accordingly, the process of S-LCA is similar to Environmental LCA (E-LCA) and can be considered as the complement of ELCA.

  One of the main advantages of GSLCAP is that since it was developed based on the LCA concept, it is suitably compatible with the supply chain logic and also LCA-based EIA methods in supply chain design and optimization models. As another advantage, GSLCAP addresses social concerns but does not cover environmental and organizational aspects; therefore, it is consistent with social issues and specifically the sustainability paradigm in the supply chain context (Pishvaee et al., 2014).

  S-LCA methodology presentes four main steps including objective and scope definition, inventory indicators determination, social impacts assessment, and interpretation. In the impact assessment phase, subcategories have been defined that are socially significant aspects or attributes. These subcategories are classified based on stakeholder and impact categories. Indeed, each stakeholder category refers to a number of social and socio-economic impact categories that are assessed according to the inventory indicators with respect to the measurement unit. A stakeholder category is a group of stakeholders who have presumably shared interests because of their identical relations with the products and their life cycle stages. Each stakeholder category refers to a number of social/socio-economic subcategories. Five main stakeholder categories and associated impact categories addressed in GSLCAP are illustrated in Table 5.6.

5.5 Conclusions

Due to the increasing concerns about the economic, environmental and social impacts of business activities, supply chain sustainability has become an active research area in the sustainability literature in recent years. On

Table 5.6 Stakeholder categories and associated impact subcategories addressed in GSLCAP (United Nations Environment Programme, 2009).

Stakeholder categories	Impact subcategories
Workers/employees	Freedom of Association and Collective Bargaining
	Child Labor
	Fair Salary
	Working Hours
	Forced Labor
	Equal opportunities/Discrimination
	Health and Safety
	Social Benefits/Social Security
Local community (national and global)	Access to material resources
	Access to immaterial resources
	Delocalization and Migration
	Cultural Heritage
	Safe & healthy living conditions
	Respect for indigenous rights
	Community engagement
	Local employment
	Secure living conditions
Society	Public commitments to sustainability issues
	Contribution to economic development
	Prevention & mitigation of armed conflicts
	Technology development
	Corruption
Consumers (including ultimate consumers or the consumers of each step of the SC)	Health & Safety
	Feedback Mechanism
	Consumer Privacy
	Transparency
	End of life responsibility
Value chain actors (excluding consumers)	Fair competition
	Promoting social responsibility
	Supplier relationships
	Respect for intellectual property rights

the other hand, the environmental impacts caused by biofuel supply chain processes such as large scale biomass feedstocks production, handling and transportation of biomass from biomass cultivation fields to conversion facilities, biomass conversion to biofuels, as well as economic impacts such as inefficient biofuel production from low energy content biomass and high costs of transportation of biomass makes the assessment of sustainability essential in the modeling and optimization of biomass-to-biofuel supply chains. This chapter investigates three pillars of sustainability paradigm, namely, economic, environmental, and social in biomass supply chains. To this aim, the economic aspect of sustainability and classification of the most common economic objective functions adopted in the biofuel supply chain design and planning models, as well as a brief description of financial ratios are provided. In order to evaluate the environmental impacts of biofuel supply chains, the four-phase framework for LCA methodology based on ISO 14040:2006 and 14044:2006 standard series is described in detail. In the same line, the list and characteristics of various life cycle impact assessment methods as well as the details of two midpoint methods, namely, CML2001 and TRACI, and two endpoint methods, namely, Eco-indicator 99 and ReCiPe which are the most credible methods of their category are described. Moreover, an LCA study is performed for a case study of a biofuel supply chain according to the EI99 method. Finally, ISO 26000:2010, which provides a framework for social responsibility implementation, as well as a brief review of SIA credible methods and guidelines, are presented.

It is noteworthy that, notwithstanding the existence of extensive literature on the environmental dimension of sustainability in the design and planning of the supply chains, few researches have addressed the social aspect of the sustainability paradigm. On the other hand, a review of the literature reveals that maximizing the number of jobs created is the most common social objective function considered as a social dimension of sustainability in the biofuel supply chain design and planning problems, which highlights the need for adopting SIA methods to quantify and optimize social impacts associated with different phases of biofuel development from construction to operation of biomass supply chains in future studies.

References

Ahi, P., & Searcy, C. (2015). Assessing sustainability in the supply chain: A triple bottom line approach. *Applied Mathematical Modelling, 39*(10), 2882–2896.
Aksoy, B., Cullinan, H., Webster, D., Gue, K., Sukumaran, S., Eden, M., & Sammons, N., Jr. (2011). Woody biomass and mill waste utilization opportunities in Alabama: Transportation cost minimization, optimum facility location, economic feasibility, and impact. *Environmental Progress & Sustainable Energy, 30*(4), 720–732.
Awudu, I., & Zhang, J. (2012). Uncertainties and sustainability concepts in biofuel supply chain management: A review. *Renewable and Sustainable Energy Reviews, 16*(2), 1359–1368.
Azadeh, A., Arani, H.V., & Dashti, H. (2014). A stochastic programming approach towards optimization of biofuel supply chain. *Energy, 76*, 513–525.
Bairamzadeh, S., Pishvaee, M. S., & Saidi-Mehrabad, M. (2015). Multiobjective robust possibilistic programming approach to sustainable bioethanol supply chain design under multiple uncertainties. *Industrial & Engineering Chemistry Research, 55*(1), 237–256.
Balaman, Ş.Y., & Selim, H. (2015). A decision model for cost effective design of biomass based green energy supply chains. *Bioresource Technology, 191*, 97–109.

Bare, J. (2011). TRACI 2.0: the tool for the reduction and assessment of chemical and other environmental impacts 2.0. *Clean Technologies and Environmental Policy*, *13*(5), 687–696. https://doi.org/10.1007/s10098-010-0338-9.

Bare, J. (2012). User's Manual-Tool for the Reduction and Assessment of Chemical and other Environmental Impacts (TRACI). United States Environmental Protection Agency: Sustainable Technology Division, Systems Analysis Branch. Retrieved from https://www.pre-sustainability.com/download/TRACI_2_1_User_Manual.pdf.

Bare, J., Norris, G. A., Pennington, D. W., & McKone, T. (2002). *TRACI-Tool for the Reduction and Assessment of Chemical and other Environmental Impacts*. Retrieved from https://www.pre-sustainability.com/download/TRACI_2_1_User_Manual.pdf.

Brand, G., Braunschweig, A., Scheidegger, A., & Schwank, O. (1998). Weighting in Ecobalances with the ecoscarcity method–Ecofactors 1997. *BUWAL (SAFEL) Environment Series, No. 297, Bern*.

Chávez, M. M. M., Sarache, W., & Costa, Y. (2018). Towards a comprehensive model of a biofuel supply chain optimization from coffee crop residues. *Transportation Research Part E: Logistics and Transportation Review*, *116*, 136–162. https://doi.org/10.1016/j.tre.2018.06.001.

Ciliberti, F., Pontrandolfo, P., & Scozzi, B. (2008). Logistics social responsibility: Standard adoption and practices in Italian companies. *International Journal of Production Economics*, *113*(1), 88–106. https://doi.org/10.1016/j.ijpe.2007.02.049.

Corporate Social Responsibility & Responsible Business Conduct. (2019). Retrieved from https://ec.europa.eu/growth/industry/corporate-social-responsibility_en.

Dale, V. H., Kline, K. L., Richard, T. L., Karlen, D. L., & Belden, W. W. (2018). Bridging biofuel sustainability indicators and ecosystem services through stakeholder engagement. *Biomass and Bioenergy*, *114*, 143–156. https://doi.org/10.1016/j.biombioe.2017.09.016.

Domac, J., Richards, K., & Risovic, S. (2005). Socio-economic drivers in implementing bioenergy projects. *Biomass and Bioenergy*, *28*(2), 97–106.

Eskandarpour, M., Dejax, P., Miemczyk, J., & Omega, O. P. (2015). *Sustainable supply chain network design: An optimization-oriented review*. Elsevier. Retrieved from https://www.sciencedirect.com/science/article/pii/S0305048315000080.

ecoinvent 3.6. (2019). Retrieved from https://www.ecoinvent.org/database/ecoinvent-36/ecoinvent-36.html.

Ethical Trading Initiative. (2018). Retrieved from https://www.ethicaltrade.org/.

Farahani, R. Z., Rezapour, S., Drezner, T., & Fallah, S. (2014). Competitive supply chain network design: An overview of classifications, models, solution techniques and applications. *Omega*, *45*, 92–118.

Fair Labor Association. (2019). Retrieved from https://www.fairlabor.org/.

GaBi. (2019). GaBi Databases 2019 Edition. Retrieved from http://www.gabi-software.com/databases/gabi-databases-2019-edition/.

Gladwin, T. N. (2001). Gladwin offers vision for sustainability. Available from: http://www.ur.umich.edu/0102/Nov05_01/10.htm.

Goedkoop, M., Hofstetter, P., Müller-Wenk, R., & Spriensma, R. (1998). The ECO-indicator 98 explained. *The International Journal of Life Cycle Assessment*, *3*(6), 352–360. https://doi.org/10.1007/BF02979347.

Gebreslassie, B. H., Yao, Y., & You, F. (2012). Design under Uncertainty of Hydrocarbon Biorefinery Supply Chains: Multiobjective Stochastic Programming Models, 28 Decomposition Algorithm, and a Comparison between CVaR and Downside Risk. *AIChE Journal*, *58*(7), 2155–2179.

Giarola, S., Bezzo, F., & Shah, N. (2013). A risk management approach to the economic and environmental strategic design of ethanol supply chains. *Biomass and Bioenergy*, *58*, 31–51.

Goedkoop, M., Heijungs, R., Huijbregts, M., Schryver, A., Struijs, J., & van Zelm, R. (2009). *ReCiPe 2008: A Life Cycle Impact Assessment Method which Comprises Harmonised Category*

Indicators at the Midpoint and the Endpoint Level, Report I: Characterisation. Netherlands: Ministry of Housing, Spatial planning and the Environment. Retrieved from https://www.rivm.nl/en/life-cycle-assessment-lca/recipe.

Goedkoop, M., Oele, M., Leijting, J., Ponsioen, T., & Meijer, E. (2016). Introduction to LCA with SimaPro Title: Introduction to LCA with SimaPro. Available from: http://www.pre-sustainability.com.

Greet. (2019). Argonne GREET Model. Available from: https://greet.es.anl.gov/.

GRI Standards. (2016). Retrieved from https://www.globalreporting.org/standards/gri-standards-download-center/.

Goedkoop, M., & Spriensma, R. (2000). The Eco-indicator 99: a damage oriented method for life cycle assessment, methodology report. *PreConsultans BV, The Netherlands, 12.*

Gonela, V., Zhang, J., Osmani, A., & Onyeaghala, R. (2015). Stochastic optimization of sustainable hybrid generation bioethanol supply chains. *Transportation research part e: Logistics and transportation review, 77,* 1–28.

Guinée, J. (2001). Handbook on life cycle assessment—operational guide to the ISO standards. *The International Journal of Life Cycle Assessment, 6*(5), 255–1255. https://doi.org/10.1007/BF02978784.

Hollander, J. (2003). The real environmental crisis: why poverty, not affluence, is the environment's number one enemy. Available from: https://books.google.com/books?hl=en&lr=&id=xKyBH-0zRRwC&oi=fnd&pg=PR10&dq=THE+REAL+ENVIRONMENTAL+CRISIS+Why+Poverty,+Not+Affluence,+Is+the+Environment's+Number+One+Enemy&ots=quV50V-FrM&sig=ucYB3eA428ZeAo5GfQa_y_RoR7g

Hauschild, M., & Potting, J. (2005). Spatial Differentiation in Life Cycle Impact Assessment–The EDIP 2003 Methodology. *Environmental News No. 80, Danish Environmental Protection Agency.*

Holme, R., & Watts, P. (1999). Corporate social responsibility: Meeting changing expectations. Conches-Geneva, Switzerland. *World Business Council for Sustainable Development.*

Huijbregts, M. A. J., Steinmann, Z. J. N., Elshout, P. M. F., Stam, G., Verones, F., Vieira, M., & van Zelm, R. (2017). ReCiPe2016: A harmonised life cycle impact assessment method at midpoint and endpoint level. *The International Journal of Life Cycle Assessment, 22*(2), 138–147. https://doi.org/10.1007/s11367-016-1246-y.

Jolliet, O., Margni, M., Charles, R., Humbert, S., Payet, J., Rebitzer, G., & Rosenbaum, R. (2003). IMPACT 2002+: A new life cycle impact assessment methodology. *The International Journal of Life Cycle Assessment, 8*(6), 324–330.

Kostin, A. M., Guillén-Gosálbez, G., Mele, F. D., Bagajewicz, M. J., & Jiménez, L. (2012). Design and planning of infrastructures for bioethanol and sugar production under demand uncertainty. *chemical Engineering Research and Design, 90*(3), 359–376.

Leduc, S., Natarajan, K., Dotzauer, E., McCallum, I., & Obersteiner, M. (2009). Optimizing biodiesel production in India.. *Applied Energy, 86,* S125–S131.

Li, Q., & Hu, G. (2014). Supply chain design under uncertainty for advanced biofuel production based on bio-oil gasification. *Energy, 74,* 576–584.

Longinidis, P., & Georgiadis, M. C. (2011). Integration of financial statement analysis in the optimal design of supply chain networks under demand uncertainty. *International Journal of Production Economics, 129*(2), 262–276.

Marufuzzaman, M., Ekşioğlu, S. D., & Hernandez, R. (2014). Environmentally friendly supply chain planning and design for biodiesel production via wastewater sludge. *Transportation Science, 48*(4), 555–574.

Marvin, W. A., Schmidt, L. D., Benjaafar, S., Tiffany, D. G., & Daoutidis, P. (2012). Economic optimization of a lignocellulosic biomass-to-ethanol supply chain. *Chemical Engineering Science, 67*(1), 68–79.

Mohseni, S., Pishvaee, M. S., & Sahebi, H. (2016). Robust design and planning of microalgae biomass-to-biodiesel supply chain: A case study in Iran. *Energy, 111,* 736–755. doi: 10.1016/j.energy.2016.06.025.

National Renewable Energy Laboratory. (2012). U.S. Life Cycle Inventory Database. Retrieved from https://www.nrel.gov/lci/.

openLCA. (2012). GreenDelta openLCA. Retrieved from http://www.openlca.org/greendelta/.

Mu, D., Seager, T., Rao, P. S., & Zhao, F. (2010). Comparative life cycle assessment of lignocellulosic ethanol production: biochemical versus thermochemical conversion. *Environmental Management, 46*(4), 565–578.

Osmani, A., & Zhang, J. (2013). Stochastic optimization of a multi-feedstock lignocellulosic-based bioethanol supply chain under multiple uncertainties. *Energy, 59*, 157–172.

Osmani, A., & Zhang, J. (2014). Optimal grid design and logistic planning for wind and biomass based renewable electricity supply chains under uncertainties. *Energy, 70*, 514–528.

Pishvaee, M. S., Razmi, J., & Torabi, S. A. (2014). An accelerated Benders decomposition algorithm for sustainable supply chain network design under uncertainty: A case study of medical needle and syringe supply chain. *Transportation Research Part E: Logistics and Transportation Review, 67*, 14–38.

Rodriguez, S.I., Roman, M.S., Sturhahn, S.C., & Terry, E.H. (2002). Sustainability Assessment and Reporting for the University of Michigan's Ann Arbor Campus.

Ross, S., Westerfield, R., & Jordan, B. (2010). *Fundamentals of corporate finance.* New York: McGraw-Hill Irwin.

SAI (2014). Social accountability 8000 (SA8000®). *SAI New York.*

Santibañez-Aguilar, J. E., González-Campos, J. B., Ponce-Ortega, J. M., Serna-González, M., & El-Halwagi, M. M. (2014). Optimal planning and site selection for distributed multiproduct biorefineries involving economic, environmental and social objectives. *Journal of Cleaner Production, 65*, 270–294.

Santibañez-Aguilar, J. E., Morales-Rodriguez, R., González-Campos, J. B., & Ponce-Ortega, J. M. (2016). Stochastic design of biorefinery supply chains considering economic and environmental objectives. *Journal of Cleaner Production, 136*, 224–245.

Sharma, B., Ingalls, R. G., Jones, C. L., Huhnke, R. L., & Khanchi, A. (2013). Scenario optimization modeling approach for design and management of biomass-to-biorefinery supply chain system. *Bioresource Technology, 150*, 163–171.

Srivastava, S. K. (2008). Network design for reverse logistics. *Omega, 36*(4), 535–548. https://doi.org/10.1016/j.omega.2006.11.012.

United Nations Environment Programme. (2009). Guidelines for social life cycle assessment of products.

UN Global Compact. (2019). The Ten Principles. Retrieved from https://www.unglobalcompact.org/what-is-gc/mission/principles.

Workplace Code of Conduct. (2019). Retrieved from https://www.fairlabor.org/sites/default/files/fla_code_of_conduct.pdf.

Steen, B. (1999). A systematic approach to environmental priority strategies in product development (EPS): Version 2000–General system characteristics. *Centre for Environmental Assessment of Products and Material Systems (CPM), Chalmers University of Technology, Gotheburg, Sweden.*

Tong, K., Gleeson, M. J., Rong, G., & You, F. (2014). Optimal design of advanced drop-in hydrocarbon biofuel supply chain integrating with existing petroleum refineries under uncertainty. *biomass and Bioenergy, 60,* 108–120.

Walther, G., Schatka, A., & Spengler, T. S. (2012). Design of regional production networks for second generation synthetic bio-fuel–A case study in Northern Germany. *European Journal of Operational Research, 218*(1), 280–292.

WCED (1987). *Our common future.* New York: Oxford University Press.

White, L., & Lee, G. J. (2009). Operational research and sustainable development: Tackling the social dimension. *European Journal of Operational Research, 193*(3), 683–692.

You, F., Tao, L., Graziano, D. J., & Snyder, S. W. (2012). Optimal design of sustainable cellulosic biofuel supply chains: multiobjective optimization coupled with life cycle assessment and input–output analysis. *AIChE Journal, 58*(4), 1157–1180.

Yue, D., You, F., & Snyder, S. W. (2014a). Biomass-to-bioenergy and biofuel supply chain optimization: Overview, key issues and challenges. *Computers & Chemical Engineering*, *66*, 36–56.

Yue, D., Slivinsky, M., Sumpter, J., & You, F. (2014b). Sustainable design and operation of cellulosic bioelectricity supply chain networks with life cycle economic, environmental, and social optimization. *Industrial & Engineering Chemistry Research*, *53*(10), 4008–4029.

Further reading

Demirbas, A. (2008). Biofuels sources, biofuel policy, biofuel economy and global biofuel projections. *Energy Conversion and Management*, *49*(8), 2106–2116.

ISO, I.O. for S. (2010). ISO 26000: 2010 Guidance on social responsibility.

ISO 14044. (2006). Environmental Management: Life Cycle Assessment; Principles and Framework.

CHAPTER 6

Uncertainty modeling approaches for biofuel supply chains

6.1 Introduction

Uncertainty is considered as one of the most crucial factors influencing the performance of the supply chain and, if ignored, can lead to suboptimal performance or even infeasible supply chain configurations (Shabani & Sowlati, 2016). Biomass supply chains are subject to various sources of uncertainty. In the upstream of the supply chain, biomass production is qualitatively and quantitatively affected by weather and climate conditions, making biomass supply vary from one season to another, from one year to another, and from one location to another (Poudel, Marufuzzaman, & Bian, 2016a). Other upstream uncertainties are related to biomass storage, biomass price, and biomass transportations. In biorefineries, there are several uncertainties such as capital and operating cost uncertainty, technology performance uncertainty, and biomass-to-biofuel conversion rate uncertainty, mainly arising from the fact that the biofuel industry is still in its infancy, and large-scale commercial biorefineries are not in place yet (Yue, You, & Snyder, 2014). Further down the supply chain, the biofuel market is surrounded by demand and price uncertainties and is sensitive to economic and financial fluctuations. In addition to the operational risks mentioned above, disruption risks originating from man-made or natural disasters (e.g., terrorist attacks, earthquakes, droughts, and floods) threaten the structure and operation of the supply chain. The 2012 drought experienced throughout the United States, for example, acknowledges the vulnerability of biofuel production to such extreme events (Langholtz et al., 2014).

Based on their scope, controllability, predictability, and propagation properties, supply chain risks and uncertainties are classified according to different typological schemes. A widely used scheme is to classify risks into three categories according to where in the supply chain they appear: (1) internal to the supply chain entities; (2) external to the supply chain entities but internal to the supply chain; and (3) external to the supply chain (Christopher & Peck, 2004). Inspired by this scheme, the risk sources of biomass supply chain are identified and classified into the above three categories, which are presented in Table 6.1. For a detailed description of this

Biomass to Biofuel Supply Chain Design and Planning under Uncertainty Copyright © 2021 Elsevier Inc.
http://dx.doi.org/10.1016/B978-0-12-820640-9.00006-4

Table 6.1 Taxonomy of uncertainty types in biomass supply chains.

Category of uncertainty sources	Uncertainty type	Subcategory
Internal risks to SC entities	• Technology performance uncertainty • Capital and operating cost uncertainty • Environmental and Social impacts uncertainty • Yield uncertainty	• Biomass yield uncertainty • Intermediate product yield uncertainty • Final product yield uncertainty
Network–related risks	• Supply-side risk	• Biomass availability uncertainty • Biomass quality uncertainty • Biomass price uncertainty • Biomass storage uncertainty
	• Transportation risks	• Supply-side transportation uncertainty • Demand-side transportation uncertainty
	• Demand-side risks	• Biofuel and by-products price uncertainty • Biofuel demand uncertainty
External risks to SC	• Environment uncertainty	• Variability of weather and climate factors (e.g., temperature, solar radiation, and precipitation) • Scarcity of resources (e.g., water and land) • Natural catastrophic events (e.g., floods, earthquakes, hurricanes, landslides and droughts)
	• Economic uncertainty	• Price volatility of raw materials (e.g., water, fertilizers, and chemicals) and utilities (e.g., electricity and gas), • Changes in oil prices, inflation, etc.
	• Socio-political uncertainty	• Government policy uncertainty • Social uncertainty • Political uncertainty

scheme and uncertainties in biofuel supply chains, the reader is referred to Chapter 4.

In choosing the most suitable approach for handling uncertainty in biofuel supply chains, one must consider the amount of data available to model uncertainty. The importance of this criterion lies in the fact that the amount of available data plays an important role in determining the type (i.e., nature) of uncertainty and its corresponding treatment approaches. When uncertainty, specifically in the context of supply chain optimization, is examined through the lens of availability of data, three types of uncertainty can be distinguished: randomness uncertainty, epistemic uncertainty, and deep uncertainty (Bairamzadeh, Saidi-Mehrabad, & Pishvaee, 2018a). Under randomness uncertainty, there is reliable historical data to exactly or approximately estimate the probability distributions quantifying the uncertain parameters of supply chain models (Klibi, Martel, & Guitouni, 2010). The uncertainty in the demand of commonly used biofuels, for example, belongs to this type of uncertainty because its probability distributions can be perfectly or ambiguously obtained with the help of available data. However, it is impossible in many real-world problems to extract probability distributions due to limited or unreliable data (Pishvaee, Rabbani, & Torabi, 2011). In such situations, subjective probability distributions can be estimated based on experts' judgments. When the opinion of experts is elicited, the epistemic uncertainty arises because of the lack of knowledge about the credibility of information provided by them (Hester, 2012). The estimates of biomass quality derived from expert knowledge, for example, are tainted with epistemic uncertainty. Under deep uncertainty, the available data is so limited that an objective or subjective probability distribution cannot be obtained, and only the bounds of uncertain parameters can be estimated (Klibi et al., 2010). The uncertainty in the yield of nonedible energy crops not previously grown on a commercial scale comes to mind as an example of deep uncertainty.

After determining the type of uncertainty, an appropriate optimization approach for coping with it must be selected. Fig. 6.1 provides an overview of leading optimization approaches for hedging against various types of uncertainty. A detailed description of these approaches along with their mathematical formulations is further described in the following.

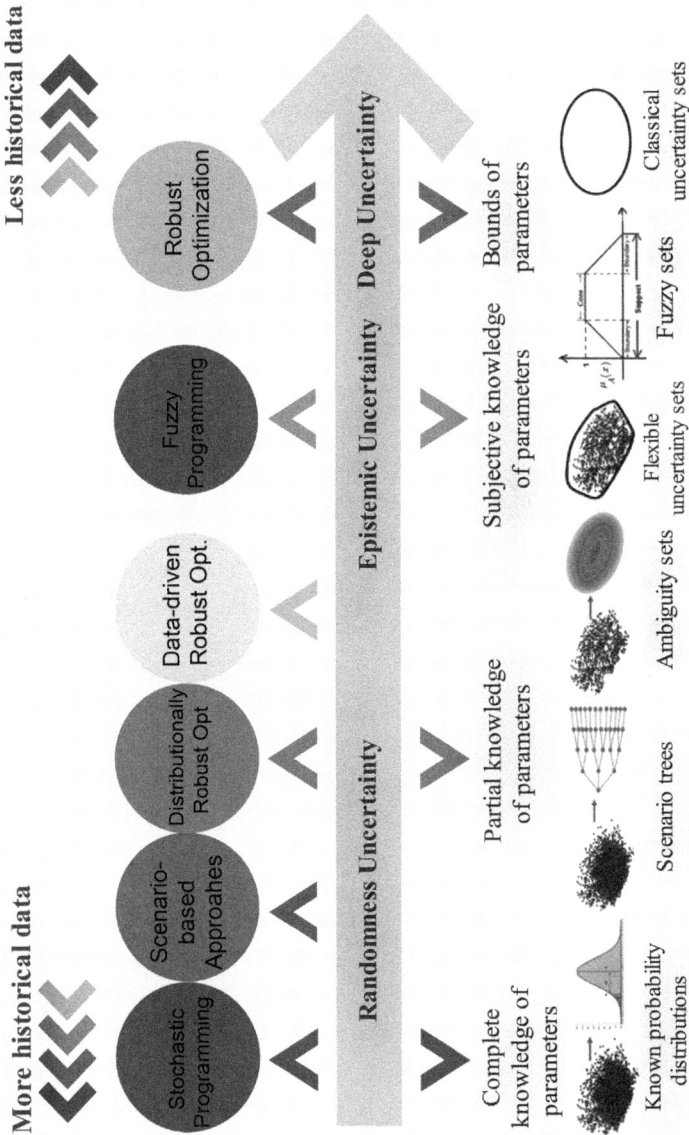

Figure 6.1 *Taxonomy of uncertainty types and modeling approaches based on the nature of uncertainty.*

6.2 Stochastic programming

As a traditional approach to handle uncertainty, stochastic programming models uncertain parameters as a multidimensional random variable that follows a known probability distribution. The goal of stochastic programming is to optimize the expected value of the objective function over all possible realizations of the random variable:

$$\inf_{x \in X} \; \mathbb{E}_{\xi}\left[f\left(x, \xi\right) \right] \tag{6.1}$$

This formulation assumes a risk–neutral attitude, meaning that potential losses in some realizations are offset by potential gains in other realizations. However, there are many situations where the decision maker has a risk-averse preference, being more concerned about large losses than average performance. To address this problem, the objective function can be defined based on different risk measures such as variance, absolute deviation, standard deviation, conditional value at risk (CVaR), central semideviation, and semideviation from a predetermined target (Govindan, Fattahi, & Keyvanshokooh, 2017).

In many real-world problems, the uncertainty is revealed sequentially, and the decision makers need to revise their decisions after observing the realizations of uncertainty. Stochastic programming with recourse has been developed to capture the sequential nature of the decision-making process. This popular class of stochastic programming is categorized into two–stage and multistage stochastic programming (Birge & Louveaux, 2011). The general formulation of a two-stage stochastic program is as follows:

$$\min_{x} \; c^T x + \mathbb{E}_{\xi}\left[f\left(x, \xi\right) \right]$$
$$s.t. \quad Ax \leq b \tag{6.2}$$

where $f\left(x, \xi\right)$ is the recourse function defined as

$$\min_{y} \; d\left(\xi\right)^T y\left(\xi\right)$$
$$s.t. \quad T\left(\xi\right)x + W\left(\xi\right)y\left(\xi\right) \leq h\left(\xi\right) \tag{6.3}$$

The first-stage decisions are those that must be decided prior to the actual realizations of the uncertain parameters, such as determining the location and capacity of biorefineries, while the second-stage decisions interpreted as recourse or corrective measures are those that can be made after the uncertain parameters are disclosed, such as the planning and scheduling of biomass-to-biofuel processing tasks. By considering that the uncertainty is uncovered sequentially in multiple stages, the two-stage problem can readily be extended to a multistage setting:

$$\min_{x} c^T x + \mathbb{E}_{\xi_1}\left[\min_{y_1 \in \mathcal{Y}_1(x,\xi_1)} d_1^T y_1 + \mathbb{E}_{\xi_2}\left[\cdots + \mathbb{E}_{\xi_T}\left[\min_{y_T \in \mathcal{Y}_T(y_{T-1},\xi_T)} d_T^T y_T\right]\right]\right] \quad (6.4)$$

where the recourse decisions $\left[y_1,\ldots,y_T\right]$ are made in stages 1 to T, and the uncertainties $\left[\xi_1,\ldots,\xi_T\right]$ are unfolded over stages 1 to T, respectively.

Employing continuous probability distributions for the random variables often leads to computationally intractable stochastic problems. Moreover, due to its complicated structures, the uncertainty data cannot easily be fitted to known continuous distributions. To overcome this problem, one can model uncertain parameters by a set of scenarios obtained by discretizing the probability distributions. Under the assumption of discrete random variables, the above stochastic formulations are simplified to the so-called scenario-based stochastic programming models.

For example, two-stage stochastic programming with discrete scenarios can be formulated as follows:

$$\min \quad c^T x + \sum_{s \in S} \pi^s (d^s)^T y^s$$

$$\text{s.t.} \quad Ax \le b \qquad\qquad\qquad\qquad (6.5)$$

$$T^s x + W^s y^s \le h^s, \quad \forall s \in S$$

where S is a set of scenarios and π^s is the probability of occurrence for each scenario $\left\{s \in S \middle| \sum_s \pi^s = 1\right\}$.

6.2.1 Chance-constrained programming

Introduced by Charnes and Cooper (1959), chance-constrained programming (CCP) or probabilistic programming is a well-known branch of stochastic programming that optimizes the objective function while controlling the confidence level of meeting constraints. The main idea of CCP is that constraints with uncertain parameters are satisfied with a specified probability to ensure a certain reliability level. In other words, CCP allows an acceptable level of constraint violation rather than enforcing constraints for all possible values of uncertain parameters that might be too costly or even unnecessary. A standard formulation of CCP is given by:

$$\min_{x}\ f(x)$$

$$s.t.\quad Prob\{G(x,\xi)\leq 0\}\geq 1-\alpha \tag{6.6}$$

where ξ is a random vector with probability distribution F, and x is the vector of decision variables that are guaranteed to satisfy the underlying constraints with a probability of more than $1-\alpha$. By changing the confidence level $\alpha \in (0,1)$, the decision maker can adjust the objective function value (i.e., economic performance) against the risk of not meeting the constraints (i.e., system reliability). Although CCP is a flexible modeling tool to incorporate uncertainty and the decision maker preference in optimization problems, it suffers from two major limitations (Nemirovski & Shapiro, 2007). First, it is difficult to calculate the probability of satisfying the constraints for a given point x because it requires the computation of multidimensional integrals. Second, the feasible region of CCP models is usually nonconvex even in the case of convex $G(x,\xi)$ with respect to the decision variables for any uncertainty realization.

6.2.2 Robust scenario-based stochastic programming

Another branch of stochastic programming is robust scenario-based stochastic programming (RSSP), which can be viewed as an improved version of scenario-based stochastic programming. Mulvey, Vanderbei, and Zenios (1995) introduced the concepts of solution robustness and feasibility robustness and then developed the RSSO model by incorporating these two robustness measures into the optimization problem. Solution robustness means the optimal solution will not change significantly among different scenarios representing possible realizations of uncertain parameters, and feasibility robustness means the optimal solution remains feasible for almost all possible scenarios. The general formulation of RSSP can be presented as follows:

$$\min\quad \sum_{s\in S}\pi^{s}Z^{s}+\lambda\sum_{s\in S}\pi^{s}\left(Z^{s}-\sum_{s'\in S}\pi^{s'}Z^{s'}\right)^{2}+\left(1-\lambda\right)\sum_{s\in S}\pi^{s}\theta^{s}$$

$$s.t.\quad Ax\leq b$$

$$Z^{s}=c^{T}x+(d^{s})^{T}y^{s},\quad \forall s\in S$$

$$T^{s}x+W^{s}y^{s}\leq h^{s}+\theta^{s},\quad \forall s\in S \tag{6.7}$$

$$\theta^{s}\geq 0$$

The objective function is composed of three terms. The first term that also appears in conventional scenario-based stochastic models is mean value, which controls the average performance. The second term controls solution robustness by measuring the deviations of the solution under different scenarios. The third term controls feasibility robustness by penalizing the violations θ^s allowed in the constraints. The weighting parameter λ, which is adjusted based on the risk attitude of the decision maker, makes a trade-off between solution and feasibility robustness. To linearize the objective function, its quadratic term is replaced with absolute deviation as follows:

$$\min \sum_{s\in S} \pi^s Z^s + \lambda \sum_{s\in S} \pi^s \left| Z^s - \sum_{s'\in S} \pi^{s'} Z^{s'} \right| + (1-\lambda)\sum_{s\in S}\pi^s\theta^s \qquad (6.8)$$

which can be transformed into an equivalent linear function by introducing auxiliary variables (Leung, Tsang, Ng, & Wu, 2007):

$$\min \quad \sum_{s\in S} \pi^s Z^s + \lambda \sum_{s\in S} \pi^s \left(Z^s - \sum_{s'\in S} \pi^{s'} Z^{s'} + 2\omega^s \right) + (1-\lambda)\sum_{s\in S}\pi^s\theta^s$$

$$s.t. \quad Z^s - \sum_{s'\in S} \pi^{s'} Z^{s'} + \omega^s \geq 0, \quad \forall s \in S$$

$$Ax \leq b \qquad (6.9)$$

$$Z^s = c^T x + (d^s)^T y^s, \qquad \forall s \in S$$

$$T^s x + W^s y^s \leq h^s + \theta^s, \qquad \forall s \in S$$

$$\theta^s \geq 0$$

6.3 Robust optimization

Although widely used to cope with uncertainty in theoretical supply chain problems, stochastic programming approaches suffer from three major drawbacks in real-world environments. First, continuous stochastic programming needs accurate knowledge about the probability distribution of random variables, which is rarely available in practice. Second, due to their discrete nature, the scenarios utilized in scenario-based approaches are very likely to not be representative of all possible values of the uncertain parameters; therefore, the solutions provided by these approaches might be infeasible for some realizations of the uncertain parameters. Third, they often employ a large set of scenarios to be responsive to different values the uncertain parameters may take, leading to computationally intractable problems (Pish-vaee et al., 2011).

As a promising alternative approach, robust optimization represents the uncertainty in parameters by a convex uncertainty set encapsulating all or almost all possible realizations of the uncertainty parameters. The goal of robust optimization is to find an optimal solution that is immunized against different realizations within the uncertainty set. In other words, robust optimization focuses on hedging against the worst-case realizations by forcing the uncertain parameters to take their worst-case values on the boundary of the uncertainty set. A general robust optimization model is defined as follows (Ben-Tal & Nemirovski, 1999):

$$\inf_{x \in X} \ \sup_{\xi \in U} f\left(x, \xi\right) \tag{6.10}$$

where x denotes the decision variables, ξ denotes the uncertain parameters, and U is an uncertainty set that specifies the range of uncertainty realizations.

Uncertainty set-induced robust optimization has become an appealing methodology to deal with uncertainty due to three main advantages. First, it only needs the lower and upper bounds of the uncertain parameters to model the uncertainty, while more detailed information is required in stochastic programming and scenario-based approaches. Second, it provides a solution that remains feasible in the face of various realizations of the uncertain parameters within the uncertainty set, while robust scenario-based stochastic optimization ensures feasibility only for those realizations that match the predefined scenarios. Third, it circumvents the computational challenges posed by stochastic programming and scenario-based approaches (Mohseni & Pishvaee, 2019).

6.3.1 Preliminaries

Let us consider the following simple linear programming problem:

$$\begin{aligned} \min_{x} \ & c^T x \\ s.t. \ & Ax \leq b \\ & x \geq 0 \end{aligned} \tag{6.11}$$

Without loss of generality, it could be assumed that only the left-hand side coefficients are tainted with uncertainty. The uncertainty in the objective function and the right-hand side coefficients can be incorporated by equivalently reformulating the original problem as follows (Li, Ding, & Floudas, 2011):

$$
\begin{aligned}
&\min_{x} \quad f\\
&s.t. \quad c^T x - f \le 0\\
&\qquad Ax + x_0 b \le 0\\
&\qquad x_0 = -1\\
&\qquad x \ge 0
\end{aligned}
\tag{6.12}
$$

Each entry $\tilde{a}_{ij}$ of matrix A can be modeled as a symmetric random variable that takes values in the interval $\left[a_{ij} - \hat{a}_{ij}, a_{ij} + \hat{a}_{ij} \right]$, where a_{ij} and $\hat{a}_{ij}$ denote the nominal value and the variation amplitude of the uncertain parameter $\tilde{a}_{ij}$, respectively. Therefore, $\tilde{a}_{ij}$ can be decomposed as

$$
\tilde{a}_{ij} = a_{ij} + \xi_{ij} \hat{a}_{ij}
\tag{6.13}
$$

where ξ_{ij} is a random variable symmetrically distributed within the interval $[-1,1]$. With the above definition, the constraint of the model (6.11) can be equivalently rewritten as follows:

$$
\sum_{j} a_{ij} x_{ij} + \sum_{j \in j_i} \xi_{ij} \hat{a}_{ij} x_{ij} \le b_i \quad \forall i
\tag{6.14}
$$

where j_i is the set of uncertain coefficients. The goal is to obtain a solution that remains feasible for all values of ξ_{ij}, which are enclosed by an uncertainty set. To ensure that the constraint is satisfied for any ξ_{ij} within the set, its robust counterpart is formulated as

$$
\sum_{j} a_{ij} x_{ij} + \max_{\xi \in U} \left[\sum_{j \in j_i} \xi_{ij} \hat{a}_{ij} x_{ij} \right] \le b_i \quad \forall i
\tag{6.15}
$$

For a given optimization problem, multiple robust versions can be obtained depending on the structure of the uncertainty set. In the robust optimization literature, different uncertainty sets have been proposed, which are presented in the following along with their corresponding robust counterpart formulations (Li et al., 2011).

6.3.1.1 Box uncertainty set
The box uncertainty set is defined using the $\infty - norm$ of the uncertainty variable ξ as follows:

$$
U_\infty = \left\{ \xi \big\| \xi_j \big| \le \Psi, \forall j \in j_i \right\}
\tag{6.16}
$$

where Ψ is a user-defined parameter that adjusts the size of the uncertainty set. The recommended range for this parameter is $[0,1]$, because when it is set to more than 1, the region of the uncertainty set exceeds the variation range of the uncertain parameters (which is corresponding to the variation range of $\xi \in [-1,1]$), leading to unreasonably conservative solutions. Replacing the inner maximization problem in the constraint (6.15) with its dual form, the robust counterpart under the box uncertainty set is formulated as follows:

$$\sum_j a_{ij} x_{ij} + \Psi \sum_{j \in J_i} \hat{a}_{ij} x_{ij} \leq b_i \quad \forall i \tag{6.17}$$

When $\Psi = 1$, the above constraint is identical to the robust formulation proposed by Soyster (1973), in which all uncertain parameters are fixed at their worst-case values. The Soyster's approach ensures the solution feasibility for all the possible realizations, but it imposes a heavy sacrifice in terms of the objective function value, which is interpreted as the cost of robustness.

6.3.1.2 Ellipsoidal uncertainty set
This uncertainty set is defined using the $2-norm$ of the uncertainty variable ξ as follows:

$$U_2 = \left\{ \xi \left| \sum_{j \in J_i} \xi_j^2 \leq \Omega^2 \right. \right\} \tag{6.18}$$

The adjustable parameter Ω must be set within the range $\left[0, \sqrt{|J_i|}\right]$, where $|J_i|$ is the cardinality of J_i (i.e., the number of the uncertain parameters in the constraint). As the value of Ω increases from 0 to $\sqrt{|J_i|}$, the size of the uncertainty set and the robustness of the solution increase. But, further increases beyond $\sqrt{|J_i|}$ are unjustifiable because the variation range of the uncertain parameters is entirely covered by the uncertainty set when $\Omega = \sqrt{|J_i|}$. Ben-Tal and Nemirovski (1999) showed that the robust counterpart under this uncertainty set is formulated as follows:

$$\sum_j a_{ij} x_{ij} + \Omega \sqrt{\sum_{j \in J_i} \hat{a}_{ij}^2 x_{ij}^2} \leq b_i \quad \forall i \tag{6.19}$$

This robust model can adjust the conservative level of solutions but it results in a nonlinear programming problem.

6.3.1.3 Polyhedral uncertainty set

This uncertainty set is defined using the $1-norm$ of the uncertainty variable ξ as follows:

$$U_1 = \left\{ \xi \middle| \sum_{j \in j_i} |\xi_j| \leq B \right\} \tag{6.20}$$

The recommended range for the user-defined parameter B is $\left[0, |J_i|\right]$. Polyhedral uncertainty set-based robust counterpart is formulated as follows (Li et al., 2011):

$$\begin{cases} \sum_j a_{ij} x_{ij} + q_i B_i \leq b_i \\ q_i \geq \hat{a}_{ij} x_{ij} \quad \forall j \in j_i \end{cases} \quad \forall i \tag{6.21}$$

Although this robust model can preserve the linearity of the initial optimization problem, it has a poor performance in adjusting the conservatism level because it finds the uncertain parameter with $\max\left(\hat{a}_{ij} x_{ij}\right)$, which has the highest effect on the constraint feasibility and regulates the conservatism level by adding $\max\left(\hat{a}_{ij} x_{ij}\right)$ multiplied by the parameter Γ.

6.3.1.4 Box-ellipsoidal uncertainty set

This uncertainty set is constructed by combining the box and ellipsoidal sets as follows:

$$U_{2 \cap \infty} = \left\{ \xi \middle| \sum_{j \in j_i} \xi_j^2 \leq \Omega^2 , \ |\xi_j| \leq 1, \forall j \in j_i \right\} \tag{6.22}$$

where Ω is an adjustable parameter that takes values in the interval $\left[1, |J_i|\right]$. When $\Omega = 1$, the box is circumscribed about the ellipsoid and the combined box-ellipsoidal uncertainty set is reduced to the ellipsoidal uncertainty set; when $\Omega = \sqrt{|J_i|}$, the box is inscribed in the ellipsoid and the combined box–ellipsoidal uncertainty set is exactly similar to the box uncertainty set. When Ω is set between these two values, the uncertainty space is bounded by both the box and ellipsoidal uncertainty sets. Fig. 6.2 illustrates this issue for the case that there are two uncertain coefficients in the constraint. The bound imposed by the box uncertainty set ensures that the uncertainty variable ξ varies within the interval $[-1,1]$ and the uncertainty

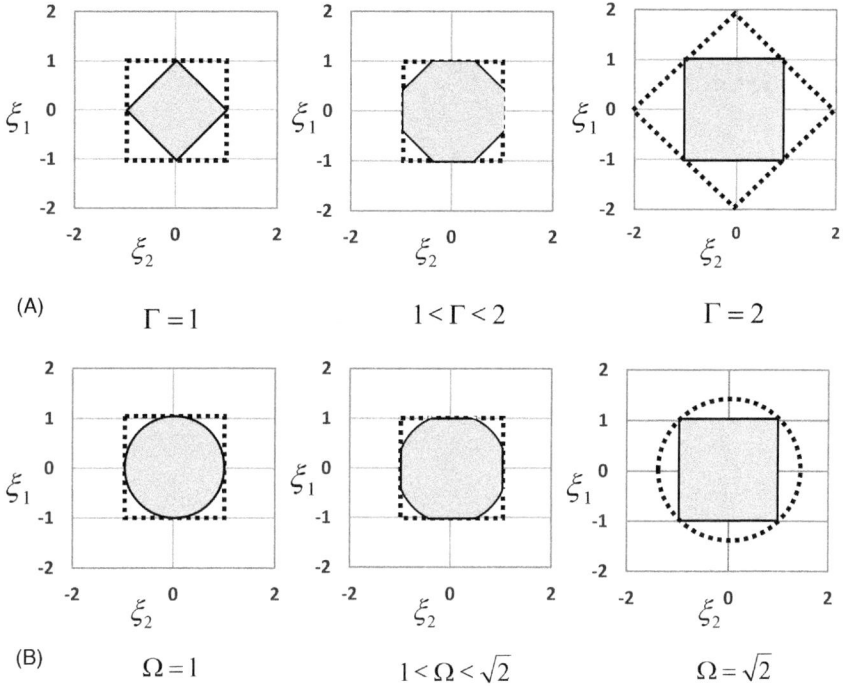

(A) $\Gamma = 1$ $1 < \Gamma < 2$ $\Gamma = 2$

(B) $\Omega = 1$ $1 < \Omega < \sqrt{2}$ $\Omega = \sqrt{2}$

Figure 6.2 Structure of (A) box-polyhedral uncertainty set and (B) box-ellipsoidal uncertainty set.

space covers only possible values of the uncertain parameters within the perturbation range, not values outside the range that may be included in the ellipsoidal uncertainty set. Under this combined uncertainty set, robust counterpart model is formulated as follows (Ben–Tal & Nemirovski, 1999):

$$\begin{cases} \sum_j a_{ij} x_{ij} + \sum_{j \in j_i} \hat{a}_{ij} u_{ij} + \Omega_i \sqrt{\sum_{j \in j_i} \hat{a}_{ij}^2 q_{ij}^2} \leq b_i \\ -u_{ij} \leq x_{ij} - q_{ij} \leq u_{ij} \quad \forall j \in j_i \end{cases} \qquad \forall i \qquad (6.23)$$

The above robust model provides a trade-off between robustness and optimality. As the value of Ω increases, the size of the uncertainty set increases and the probability of constraint violation decreases, meaning that the solution becomes more robust against data variations, but at the same time, the objective function deterioration (i.e., the cost of robustness) increases. A practical drawback of this model is that it destroys the linearity of the original problem.

6.3.1.5 Box-polyhedral uncertainty set

This uncertainty set as the intersection of the box and polyhedral uncertainty sets is described as follows:

$$U_{1 \cap \infty} = \left\{ \xi \left| \sum_{j \in j_i} \left| \xi_j \right| \leq B, \, \left| \xi_j \right| \leq 1, \forall j \in j_i \right. \right\} \qquad (6.24)$$

where the user-defined parameter B is tuned in the range $\left[1, \left| j_i \right| \right]$, making sure the combined box-polyhedral uncertainty set is not reduced to the box or polyhedral set. Fig. 6.2 illustrates the structure of this uncertainty set for a constraint with two uncertain coefficients. Under this combined uncertainty set, the inner maximization problem in the constraint (6.15) is replaced with its conic dual to obtain the following robust counterpart (Li et al., 2011):

$$\begin{cases} \sum_j a_{ij} x_{ij} + q_i B_i + \sum_{j \in j_i} p_{ij} \leq b_i \\ q_i + p_{ij} \geq \hat{a}_{ij} x_{ij} \quad \forall j \in j_i \end{cases} \quad \forall i \qquad (6.25)$$

This robust model not only can adjust the conservatism level of the solution against the loss in the objective function value but also can keep the linearity of the original problem.

The above robust counterpart model has been derived in a different manner by Bertsimas and Sim (2004). Instead of confining the possible realizations by an uncertainty set, they introduced a parameter Γ, called the uncertainty budget, to control the number of uncertain parameters getting their worst-case values. With this purpose, they proposed the following constraint:

$$\sum_j a_{ij} x_{ij} + \max_{\left\{ S_i \cup t_i \mid S_i \subset j_i, \, S_i = \lfloor \Gamma_i \rfloor, t_i \in j_i \setminus S_i \right\}} \left\{ \sum_{j \in S_i} \hat{a}_{ij} x_{ij} + \left(\Gamma_i - \lfloor \Gamma_i \rfloor \right) \hat{a}_{it_i} x_{it_i} \right\} \leq b_i \quad \forall i \qquad (6.26)$$

where s_i is the subset that incorporates $\lfloor \Gamma_i \rfloor$ uncertain coefficients allowed to reach their worst-case values, and t_i denotes an additional coefficient that changes by $\left(\Gamma_i - \lfloor \Gamma_i \rfloor \right) \hat{a}_{it_i}$ if Γ_i is chosen as an integer. Given an optimal solution x^*, the protection function of the above constraint (i.e., the inner maximization) is equivalent to

$$\max \quad \sum_{j \in j_j} \hat{a}_{ij} x_{ij}^{\star} z_{ij}$$

$$s.t. \quad \sum_{j \in j_j} z_{ij} \leq \Gamma_i \qquad (6.27)$$

$$0 \leq z_{ij} \leq 1 \quad \forall j \in j_i$$

Substituting the dual problem of (6.27) into (6.26), the following robust counterpart is obtained:

$$\begin{cases} \sum_j a_{ij} x_{ij} + q_i \Gamma_i + \sum_{j \in j_i} p_{ij} \leq b_i \\ q_i + p_{ij} \geq \hat{a}_{ij} x_{ij} \quad \forall j \in j_i \end{cases} \quad \forall i \qquad (6.28)$$

6.3.2 Adjustable robust optimization

Conventional set-based robust optimization approaches are classified as static robust optimization that assumes all decision variables are here-and-now, meaning that they must be determined before observing the actual realization of the uncertain parameters. This assumption, however, is very unrealistic in many dynamic real-world environments where some decisions can be made or revised when part of the uncertainty becomes known. Take for instance a biomass supply chain design problem in which the location of biorefineries needs to be decided at time zero, while the decision of how much biofuel produced can be made in each time stage based on the realizations of demand in the previous stages. To capture the sequential nature of decision-making in such problems, Ben-Tal, Goryashko, Guslitzer, and Nemirovski (2004) introduced adjustable robust optimization (ARO) that allows the recourse decisions to adjust themselves to the corresponding revealed information. Because of its flexibility in tuning the decision variables, ARO is generally less conservative than the traditional robust optimization, but this flexibility comes at the expense of computational tractability. The ARO problem can be written as follows (Bertsimas & Thiele, 2006):

$$\min_{x, y(\xi)} \quad c^T x$$

$$s.t. \quad Ax \leq b \qquad (6.29)$$

$$T(\xi) x + W\gamma(\xi) \leq h(\xi) \quad \forall \xi \in U$$

where U is a convex uncertainty set in which the uncertainty realizations reside, T and h are the uncertain coefficients depending on the realizations, x is the here-and-now decision that made before the uncertain coefficients reveal themselves, and y is the wait-and-see decision that can be adjusted after observing the realizations. In most of the ARO literature, W is assumed constant to satisfy the fixed recourse property and to derive tractable counterparts. It is worth noting that the objective function with uncertain coefficients and here-and-now decisions can be converted into a constraint using an epigraph formulation.

The mapping function $y(\cdot)$, also called decision rule, could take any form, and as a result, ARO involves optimizing over an infinitely large set of all functions, which is computationally intractable. To yield a tractable approximation for this problem, a common approach consists in restricting the function $y(\cdot)$ to be an affine decision rule of the realized data:

$$y(\xi) = p + Q\xi \tag{6.30}$$

where p and Q are new decision variables that are to be optimized by the model. For an exhaustive description of different classes of linear and nonlinear decision rules, the reader is referred to Yanikoğlu, Gorissen, and Den Hertog (2019). Substituting the above decision rule into the model (6.29) results in the following Affinely Adjustable Robust Counterpart (AARC):

$$\begin{aligned} \min_{x,p,Q} \quad & c^T x \\ \text{s.t.} \quad & Ax \le b \\ & T(\xi)x + W(p + Q\xi) \le h(\xi) \quad \forall \xi \in U \end{aligned} \tag{6.31}$$

This model is now a standard robust optimization problem and therefore its equivalent robust counterpart can be obtained for special structures of the uncertainty set U using similar derivation procedure as that for the conventional robust optimization. The resulting problem provides the optimal value $x^\star$ that can directly be implemented as the here-and-now decisions, and the optimal values $p^\star$ and $Q^\star$ that are used for computing the wait-and-see decisions $y(\xi)$ once the uncertain parameter ξ is recovered.

6.4 Data-driven optimization

Stochastic programming and robust optimization have been the dominant mathematical programming approaches for decision-making under uncertainty, each having its own advantages and disadvantages. One important drawback of robust optimization is that it usually yields costly over-conservatism solutions. There are two main reasons for this. First, robust optimization focuses only on the support of the uncertain parameters without incorporating any information about their distributional characteristics (Rahimian & Mehrotra, 2019). Second, classical uncertainty sets adopted in robust optimization have a fixed structure, lacking sufficient flexibility to be adjusted according to the characteristics of the underlying uncertainty (Mohseni & Pishvaee, 2019). On the other hand, stochastic programming typically optimizes the expected performance of the solution, but it entails the availability of the complete knowledge about the probability distributions of uncertain parameters (Pishvaee et al., 2011). As an intermediate approach between robust optimization and stochastic programming, data-driven optimization constructs flexible or ambiguous uncertainty sets from the uncertainty data, thereby enabling the injection of information from available historical data into the mathematical programming problems. Data-driven robust optimization and distributionally robust optimization are two main approaches for data-driven optimization under uncertainty, which are described in the following sections.

6.4.1 Data-driven robust optimization

Although conventional robust optimization approaches are powerful tools to cope with uncertainty, two important difficulties appear in their practical use. First, the classical uncertainty sets are defined and fixed a priori, rather than being constructed based on information extracted from historical data. This limited flexibility forces the user to select the type of uncertainty set from those previously developed, and in most cases, the chosen uncertainty set is not able to capture the intrinsic complexity of uncertainty data. For example, the correlation and asymmetry embodied in data samples displayed in Fig. 6.3 cannot be effectively handled by the classical uncertainty sets because these sets lead to superfluous coverage, which can lead to excessively conservative solutions with high robustness cost (Shang, Huang, & You, 2017). Second, it is challenging to tune the variation range of uncertain parameters around their nominal values, particularly in the case of high-dimensional uncertainty data. Commonly, the lower and upper bounds of the variation range are set based on the minimum and maximum values of uncertain parameters. Although this strategy ensures the coverage of all possible realizations, it can result in large-sized uncertainty sets that yield overly conservative solutions (Mohseni & Pishvaee, 2019). As shown in Fig. 6.3, smaller variation ranges can scale down the uncertainty set, but

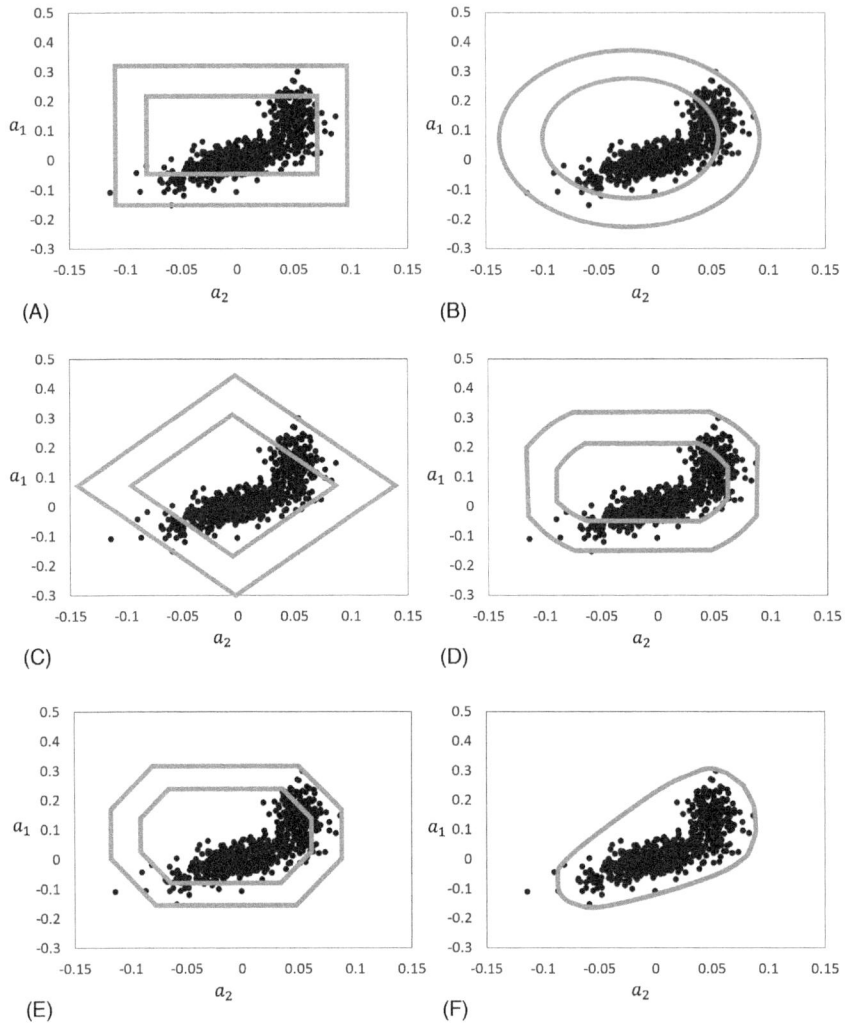

Figure 6.3 *Performance of different uncertainty sets in covering the uncertainty data.* (A) box uncertainty set, (B) ellipsoidal uncertainty set, (C) polyhedral uncertainty set, (D) box-ellipsoidal uncertainty set, (E) box-polyhedral uncertainty set, and (F) flexible uncertainty set.

this increases the number of realizations falling outside the uncertainty set and reduces the protection of the solution against infeasibility.

6.4.1.1 Flexible uncertainty set

Data–driven robust optimization (DDRO) constructs a flexible uncertainty set that accurately covers the region in which the past realizations of uncertainty data reside. The developed uncertainty set by DDRO is no longer restricted to the geometric structure of the classical uncertainty sets. As represented in Fig. 6.3, there is no void space in the flexible uncertainty set, meaning that it can avoid unnecessary conservatism associated with the classical uncertainty sets. The limitation, however, is that the performance of DDRO is heavily dependent on the amount of available data, and data samples used as inputs to the DDRO model should be a good representative of the whole uncertainty domain to ensure the reliability of the obtained solutions.

In recent years, various techniques have been proposed to form flexible uncertainty sets based on available historical data, such as copula function and cutting planes (Zhang, Jin, Feng, & Rong, 2018), the Dirichlet process mixture model (Ning & You, 2017), and support vector clustering (SVC) (Shang et al., 2017). For data-driven construction of the uncertainty set when facing high-dimensional uncertainty data, SVC is a promising approach as it can effectively capture the distributional geometry of the underlying data. Moreover, the uncertainty set constructed by SVC can preserve the linearity of the original problem when formulating the robust counterpart problem. The framework of SVC-based robust optimization proposed by Shang et al. (2017) is described in the following.

Assume that there is a set of N data samples $\{u^i\}_{i=1}^N$, each of which is a n-dimensional vector $u = \left[u_1, u_2, \ldots, u_n\right]^T$. With the help of a nonlinear mapping function Φ from the input data into a high-dimensional feature space, SVC looks for the smallest sphere encapsulating all data by the following optimization problem (Ben-Hur, Horn, Siegelmann, & Vapnik, 2001):

$$\min \ R^2 + P\sum_{i=1}^N \xi_i$$
$$s.t. \ \left\|\Phi(u^i) - C\right\|_2^2 \leq R^2 + \xi_i, \quad i = 1, \ldots, N,$$
$$\xi \geq 0$$

(6.32)

where R is the radius of the sphere centered at C, and P is a constant to penalize the violations of the data points outside the sphere. To solve this problem, the Lagrangian is introduced as follows:

$$L = R^2 + P\sum_{i=1}^N \xi_i - \sum_{i=1}^N \alpha_i \left(R^2 + \xi_i - \left\|\Phi(u^i) - C\right\|_2^2\right) - \sum_{i=1}^N \beta_i \xi_i \quad (6.33)$$

where $\alpha_i, \beta_i \geq 0$ are the Lagrangian multipliers. Taking the derivative of L with respect to R, C and ξ_i, and setting them to zero yields:

$$\sum_{i=1}^N \alpha_i = 1 \qquad (6.34)$$

$$\sum_{i=1}^{N} \alpha_i \Phi(u^i) = C \tag{6.35}$$

$$\alpha_i + \beta_i = P \tag{6.36}$$

According to the KKT complementarity conditions, the following relations are obtained:

$$\beta_i \xi_i = 0 \tag{6.37}$$

$$\alpha_i \left(R^2 + \xi_i - \|\Phi(u^i) - C\|_2^2 \right) = 0 \tag{6.38}$$

Using the above equations, the position of each data point relative to the sphere boundary can be detected. A data point with $\xi_i > 0$ that lies outside the sphere has $\beta_i = 0$ and $\alpha_i = P$ based on Eqs. (6.37) and (6.36), respectively. Such a point is named a bounded support vector (BSV). A data point with $\xi_i = 0$ resides in or on the surface of the sphere. If $0 < \alpha_i < P$, it lies exactly on the surface of the sphere based on Eq. (6.38), and is referred to as a support vector (SV). Finally, all other points that have $\alpha_i = 0$ are enclosed within the sphere and named inlier data points (ID) (Ben-Hur et al., 2001).

With the above definition, Eqs. (6.34)-(6.36) are plugged into the Lagrangian, leading to the following quadratic programming as the Wolfe dual form:

$$\max \quad \sum_{i=1}^{N} \alpha_i K(u^i, u^i) - \sum_{i=1}^{N} \sum_{j=1}^{N} \alpha_i \alpha_j K(u^i, u^j)$$

$$s.t. \quad \sum_{i=1}^{N} \alpha_i = 1 \tag{6.39}$$

$$0 \leq \alpha_i \leq P$$

where $K(u^i, u^j)$ is the kernel function representing the dot products $\Phi(u^i) \cdot \Phi(u^j)$. Using the optimal value of the above model, the radius of the sphere is calculated:

$$R^2 = \|\Phi(u^{i'}) - C\|_2^2 \quad i' \in SV$$

$$= K(u^{i'}, u^{i'}) - 2\sum_{i=1}^{N} \alpha_i K(u^{i'}, u^i) + \sum_{i=1}^{N} \sum_{j=1}^{N} \alpha_i \alpha_j K(u^i, u^j) \quad i' \in SV \tag{6.40}$$

The uncertainty set that encloses the region inside the sphere is described as follows:

$$U(u) = \left\{ u \middle| K(u,u) - 2\sum_{i=1}^{N}\alpha_i K(u,u^i) + \sum_{i=1}^{N}\sum_{j=1}^{N}\alpha_i \alpha_j K(u^i,u^j) \le R^2 \right\} \quad (6.41)$$

By substituting the radius R (6.40) into the uncertainty set (6.41), the uncertainty set is rewritten as follows:

$$U(u) = \left\{ u \middle| K(u,u) - 2\sum_{i\notin ID}\alpha_i K(u,u^i) \le K(u^{i'},u^{i'}) - 2\sum_{i\notin ID}\alpha_i K(u^{i'},u^i) \quad i' \in SV \right\}$$
$$(6.42)$$

Note that the summation is over $i \notin ID$ because $\alpha_i = 0$ for $i \in ID$. The use of the common kernel functions such as radial, polynomial, and sigmoid in the above uncertainty set leads to intractable robust counterparts. To address this problem, Shang et al. (2017) proposed a weighted generalized intersection kernel as follows:

$$K(u,v) = \sum_{k=1}^{n} l_k - \|Q(u-v)\|_1 \quad (6.43)$$

where Q is the weighting matrix calculated based on the estimated covariance matrix (6.44), and l_k is the width parameters tuned according to the relation (6.45) to retain the positive-definiteness of the kernel matrix.

$$Q = \Sigma^{-\frac{1}{2}}$$
$$= \left[\frac{1}{N-1} \left(\sum_{i=1}^{N} u^i (u^i)^T - \left(\sum_{i=1}^{N} u^i\right)\left(\sum_{i=1}^{N} u^i\right)^T \right) \right]^{-\frac{1}{2}} \quad (6.44)$$

$$l_k > \max_{i=1}^{N} q_k^T u^i - \min_{i=1}^{N} q_k^T u^i \quad (6.45)$$

where q_k is the kth column of the weighting matrix.

Plugging the kernel function (6.43) into the initial uncertainty set (6.42) and removing the constant term appearing on both side of the inequality leads to the following uncertainty set:

$$U(u) = \left\{ u \middle| \sum_{i\notin ID}\alpha_i \|Q(u-u^i)\|_1 \le \sum_{i\notin ID}\alpha_i \|Q(u^{i'}-u^i)\|_1 \quad i' \in SV \right\} \quad (6.46)$$

By defining $\Omega = \sum_{i\notin ID}\alpha_i \left\|Q(u^{i'} - u^i)\right\|_1, i' \in SV$, and introducing n-dimen-

sional auxiliary variables z_i, the uncertainty set is equivalently rewritten as:

$$U(u) = \left\{ u \left| \begin{array}{c} \exists z_i, i \notin ID, \ s.t. \\ \sum_{i\notin ID}\alpha_i \cdot z_i^T 1 \leq \Omega \\ -z_i \leq Q(u - u^i) \leq z_i, i \notin ID \end{array} \right. \right\} \tag{6.47}$$

It should be noted that the conservatism level of the uncertainty set can be controlled by the penalty parameter P used in the SVC model. As the value of P increases, the size of the sphere enclosing data samples becomes larger, and fewer samples are allowed to reside outside the sphere, resulting in a more conservative uncertainty set.

6.4.1.2 Robust counterpart formulation

Under the data-driven uncertainty set (6.47), the robust counterpart of a linear programming problem is formulated as follows:

$$\begin{array}{ll} \min_{x} & c^T x \\ s.t. & \max_{\tilde{a}\in U(u)} \tilde{a}^T x \leq b \\ & x \geq 0 \end{array} \tag{6.48}$$

The inner optimization problem in the constraint can be written as

$$\begin{array}{ll} \max_{u,z_i} & u^T x \\ s.t. & \sum_{i\notin ID}\alpha_i \cdot 1^T z_i \leq \Omega \\ & -z_i \leq Q(u - u^i) \leq z_i, \quad \forall i \notin ID \end{array} \tag{6.49}$$

By transforming the maximization problem (6.49) into its dual problem, and then incorporating the dual problem into the original constraint in the model (6.48), the following robust counterpart is obtained:

$$\min_{m_i,n_i,\lambda,x} \quad c^T x$$

$$s.t. \quad \sum_{i \notin ID}(m_i - n_i)^T Qu^i + \lambda\Omega \le b$$

$$\sum_{i \notin ID} Q\left(n_i - m_i\right) + x = 0 \qquad (6.50)$$

$$m_i + n_i = \lambda \cdot \alpha_i \cdot 1 \quad \forall i \notin ID$$

$$n_i, m_i \in R_+^n \quad x, \lambda \ge 0$$

where m_i, n_i, and λ are the dual variables introduced for formulating the dual problem.

6.4.2 Distributionally robust optimization

As stated before, in stochastic programming, it is assumed that there is perfect knowledge of the probability distributions of the uncertainty data. In practice, however, such precise information is rarely available. Moreover, the uncertainty data often exhibit complicated distributional characteristics, which make it impossible to fit into commonly known probability distributions. While stochastic programming relies on a single accurate probability distribution, distributionally robust optimization (DRO) assumes that the probability distribution governing uncertain parameters is unknown but belongs to an ambiguity set of probability distributions (Delage & Ye, 2010). The advantages of DRO can be summarized in three aspects. First, the DRO approach is well-suited to the limited historical data available in real-life problems because it models uncertainty by employing the ambiguity set that can be constructed even using partial distributional information (Rahimian & Mehrotra, 2019). Second, it improves the reliability of the solutions and avoids overly conservative solutions by incorporating available distributional information, which is ignored by robust optimization (Ning & You, 2019). Third, it inherits two salient features from robust optimization, including preserving the computational tractability of the deterministic problem and immunizing the solution against infeasibility.

6.4.2.1 Ambiguity set
The general form of DRO can be expressed as follows (Delage & Ye, 2010):

$$\inf_{x \in X} \sup_{P \in D} \mathbb{E}_p\left[f\left(x,\xi\right)\right] \qquad (6.51)$$

where ξ is a random vector with probability distribution P, and $\mathcal{D}$ is a family of probability distributions, called the ambiguity set, that is assumed to contain the true distribution P. The idea of DRO is to optimize the worst-case expectation of the objective function over all possible distributions in the ambiguity set. If the distributions residing in the ambiguity set places all their weight on the support of the random vector (i.e., the boundary of the uncertainty set), DRO reduces to robust optimization. On the other hand, if the ambiguity set consists only of the true distribution of the random vector, DRO reduces to stochastic programming (Rahimian & Mehrotra, 2019).

Commonly, the ambiguity set is constructed by moment-based approaches that define a set of all probability distributions whose moments, usually the first two moments (i.e., mean and covariance), satisfy certain conditions. Using the first and second moment information extracted from historical data, the ambiguity set is defined as follows (Yue, Chen, & Wang, 2006):

$$
\mathcal{D} = \left\{ P \in \mathcal{M}_+ \;\middle|\; \begin{array}{l} \mathbb{E}_P\{\xi\} = \mu \\[6pt] \mathbb{E}_P\left[(\xi - \mu)(\xi - \mu)^T\right] = \Sigma \end{array} \right\} \tag{6.52}
$$

where μ is the mean vector, Σ is the covariance matrix, and $\mathcal{M}_+$ represents the set of all probability distributions. Due to the fact that the mean and covariance matrix may be subject to estimation errors, Delage and Ye (2010) proposed the following generalized ambiguity set that accommodates the support of the distribution and a confidence region for the mean and the covariance matrix:

$$
\mathcal{D} = \left\{ P_\xi \in \mathcal{M}_+ \;\middle|\; \begin{array}{l} P\{\xi \in \Xi\} = 1 \\[6pt] \left(\mathbb{E}_P\{\xi\} - \mu\right)^T \Sigma^{-1} \left(\mathbb{E}_P\{\xi\} - \mu\right) \leq \gamma_1 \\[6pt] \mathbb{E}_P\left[(\xi - \mu)(\xi - \mu)^T\right] \leq \gamma_2 \Sigma \end{array} \right\} \tag{6.53}
$$

where the first constraint forces the uncertainty realizations to be confined within the support set Ξ. The adjustable parameters γ_1 and γ_2 control the size of the ambiguity set.

Another class of ambiguity set was developed by Wiesemann, Kuhn, and Sim (2014) as follows:

$$
\mathcal{D} = \left\{ P_\xi \in \mathcal{M}_+ \middle| \begin{array}{l} P\{\xi \in \Xi\} = 1 \\[2mm] \mathbb{E}_P\{f_i(\xi)\} \le \gamma_i, \ i = 1,...,k \end{array} \right\} \tag{6.54}
$$

where the second constraint incorporates moment information by k functions $\{f_i(\cdot)\}$ whose expectation is confined by a threshold γ_i. By introducing the auxiliary random variable μ, the ambiguity set can equivalently be expressed as the following set, termed as the lifted ambiguity set:

$$
\bar{\mathcal{D}} = \left\{ Q_{\xi,\mu} \in \mathcal{M}_+ \middle| \begin{array}{l} P\{(\xi,\mu) \in \bar{\Xi}\} = 1 \\[2mm] \mathbb{E}_Q\{\mu\} \le \gamma \end{array} \right\} \tag{6.55}
$$

where $\bar{\Xi}$ is the lifted support set defined as follows:

$$
\bar{\Xi} = \left\{ (\xi,\mu) \middle| \begin{array}{l} \xi \in \Xi \\[2mm] f_i(\xi) \le \mu_i, \ i = 1,...,k \end{array} \right\} \tag{6.56}
$$

Although robust counterpart formulations induced by the above ambiguity sets can generally be cast as convex optimization problems, they are computationally demanding, particularly for large-scale problems. In order to get a more computationally tractable form, the moment function $\{f_i(\cdot)\}$ can be defined as a piecewise linear function (Bertsimas, Sim, & Zhang, 2017). Adopting this idea, Shang and You (2018) reformulate the lifted support set by employing a set of linear inequalities:

$$
\bar{\Xi} = \left\{ (\xi,\mu) \middle| \begin{array}{l} \xi \le \xi^{max} \\[1mm] \xi^{min} \le \xi \\[1mm] 0 \le \mu_i \quad\ \ i = 1,...,k \\[1mm] a_i^T \xi - b_i \le \mu_i, \ i = 1,...,k \end{array} \right\} \tag{6.57}
$$

which can be expressed in the following matrix form:

$$
\bar{\Xi} = \{(\xi,\mu) | C\xi + D\mu \le e\} \tag{6.58}
$$

With $A^T = \begin{bmatrix} a_1\, a_2\, ...a_k \end{bmatrix}$ and $b = \begin{bmatrix} b_1\, b_2\, ...b_k \end{bmatrix}$, the matrices C, D and the vector e are given by:

$$C = \begin{bmatrix} I \\ I \\ 0 \\ A^T \end{bmatrix}, \quad D = \begin{bmatrix} 0 \\ 0 \\ -I \\ -I \end{bmatrix}, \quad e = \begin{bmatrix} \xi^{max} \\ -\xi^{min} \\ 0 \\ b \end{bmatrix} \tag{6.59}$$

One of the main challenges faced by DRO approaches is how to tune the parameters of the ambiguity set in such a way that the moment information of uncertainty data can effectively be captured. To address this problem, Shang and You (2018) develop a two-stage procedure to calibrate the set parameters. In the first stage, the vectors $\{a_i\}$, which are interpreted as projection directions, are determined by principal component analysis (PCA). Then, in the second stage, a number of truncation points $\{b_i\}$ are symmetrically set along each direction around the sample mean.

6.4.2.2 Robust counterpart formulation

To obtain the robust counterpart of the DRO problem under the ambiguity set (6.55) and lifted support set (6.58), consider a general two-stage DRO problem as follows:

$$\min_{x} \quad c^T x + \max_{P \in \bar{D}} E_p \left[f(x, \xi) \right] \tag{6.60}$$
$$s.t. \quad Ax \leq b$$

where

$$f(x, \xi) = \begin{cases} \min & d^T y \\ s.t. & T(\xi)x + Wy \leq h(\xi) \end{cases} \tag{6.61}$$

To remove the inner optimization problem in the objective function, it is first expressed as follows (Shang & You, 2018):

$$\sup_{Q} \int_{\Xi} p(\xi, \mu) f(x, \xi) d\xi d\mu$$
$$s.t. \int_{\Xi} p(\xi, \mu) d\xi d\mu = 1 \tag{6.62}$$
$$\int_{\Xi} p(\xi, \mu) d\xi d\mu \leq \gamma$$

Introducing the Lagrangian multipliers λ and v, the Lagrangian is written as:

$$L = \int_{\bar{\Xi}} p(\xi,\mu) f(x,\xi) d\xi d\mu - \lambda \left(\int_{\bar{\Xi}} p(\xi,\mu) d\xi d\mu - 1 \right) - v^T \left(\int_{\bar{\Xi}} p(\xi,\mu) d\xi d\mu - \gamma \right)$$

$$= \int_{\bar{\Xi}} p(\xi,\mu) \left(f(x,\xi) - \lambda - v^T \mu \right) d\xi d\mu + \lambda + v^T \gamma \qquad (6.63)$$

Relying on the strong duality theorem, Shang and You (2018) show that problem (6.62) can be reformulated as the following minimization problem:

$$\begin{aligned} &\min_{\lambda,v,x} \quad \lambda + v^T \gamma \\ &\text{s.t.} \quad \lambda + v^T \mu \geq f(x,\xi), \quad \forall (\xi,\mu) \in \bar{\Xi} \qquad (6.64) \\ &\quad\quad v \geq 0 \end{aligned}$$

where the first constraint is imposed to ensure the Lagrange dual function is bounded. Substituting the above problem into the original problem (6.60) leads to the following equivalent DRO problem:

$$\begin{aligned} &\min \quad C^T x + \lambda + v^T \gamma \\ &\text{s.t.} \quad \lambda + v^T \mu \geq d^T y(\xi,\mu), \quad \forall (\xi,\mu) \in \bar{\Xi} \\ &\quad\quad T(\xi) x + W y(\xi,\mu) \leq h(\xi), \quad \forall (\xi,\mu) \in \bar{\Xi} \qquad (6.65) \\ &\quad\quad Ax \leq b \\ &\quad\quad v \geq 0 \end{aligned}$$

Similar to what was done for adjustable robust optimization, the uncertain coefficients and the recourse decisions are assumed to be affinely dependent on their corresponding uncertainties:

$$\begin{aligned} T(\xi) &= T_0 + T_\xi \xi \\ h(\xi) &= h_0 + h_\xi \xi \qquad (6.66) \\ y(\xi,\mu) &= y_0 + Y_\xi \xi + Y_\mu \mu \end{aligned}$$

With the above definition, problem (6.65) is rewritten as follows:

$$\min \quad c^T x + \lambda + v^T \gamma$$

$$s.t. \quad \lambda + v^T \mu - d^T \gamma_0 \geq \max_{(\xi,\mu)\in\Xi} d^T \left(Y_\xi \xi + Y_\mu \mu \right)$$

$$\max_{(\xi,\mu)\in\Xi} \left(T_\xi \xi x + W \left(Y_\xi \xi + Y_\mu \mu \right) - h_\xi \xi \right) \leq h_0 - W \gamma_0 - T_0 x \quad (6.67)$$

$$Ax \leq b$$

$$v \geq 0$$

By replacing the inner maximization problems with their corresponding dual problems, the robust counterpart of the original DRO problem is obtained as follows:

$$\min \quad c^T x + \lambda + v^T \gamma$$

$$s.t. \quad \lambda - d^T \gamma_0 - M^T e \geq 0$$

$$M^T C = d^T Y_\xi$$

$$M^T D = d^T Y_\mu - v$$

$$(T_0)^T x + W^T \gamma_0 - h_0 + N^T e \leq 0 \quad (6.68)$$

$$N^T C = (T_\xi)^T x + W^T Y_\xi - h_\xi$$

$$N^T D = W^T Y_\mu$$

$$Ax \leq b$$

$$v \geq 0, N \geq 0, M \geq 0$$

6.5 Fuzzy mathematical programming

Fuzzy mathematical programming is traditionally classified into two main categories: (1) flexible programming and (2) possibilistic programming. Flexible programming addresses vagueness in the given target values of objectives and/or the flexibility in constraints satisfaction, while possibilistic programming deals with ambiguous (i.e., imprecise) parameters in the objective functions and constraints due to the lack of knowledge about the precise values of these parameters (i.e., epistemic uncertainty). Alongside the conventional categories of fuzzy mathematical programming models (i.e., the flexible and possibilistic programming), two other categories of modeling approaches, namely, fuzzy stochastic programming and robust fuzzy programming, have been proposed in the literature in the recent decade. Fuzzy stochastic programming, also known as stochastic fuzzy programming (see for instance, Van Der Vaart, De Vries, & Wijngaard, 1996; Chakraborty, 2002) deals with both types of probabilistic randomness and possibilistic imprecision of input data. Robust fuzzy programming approach, including robust possibilistic programming (Pishvaee, Razmi, & Torabi, 2012) and robust flexible programming (Pishvaee & Khalaf, 2016), extends the robust optimization scope into the fuzzy mathematical programming. Notably, robust fuzzy programming has significant benefits compared to conventional fuzzy programming approaches (Naderi, Pishvaee, & Torabi, 2016; Pishvaee et al., 2012).

6.5.1 Possibilistic programming

Possibilistic programming deals with the epistemic uncertainty of the model parameters, arising from the lack of knowledge about the precise values of these parameters. In the possibilistic programming, a possibility distribution is considered for each uncertain parameter, which displays the occurrence degree of all possible values in the corresponding interval. The possibility distribution can be estimated based on some available objective data and subjective knowledge of decision makers (DMs) in the form of fuzzy numbers, such as triangular and trapezoidal fuzzy numbers (Naderi et al., 2016; Pishvaee et al., 2012). Among the possibilistic programming approaches, the possibilistic chance constrained approach (PCCP) is described here to deal with the epistemic uncertainty of input parameters.

Let us consider the following simple linear programming problem:

$$
\begin{aligned}
\min \quad & \tilde{c}x \\
s.t. \quad & \tilde{a}x \leq b \\
& \tilde{a}x \geq \tilde{b} \\
& \tilde{a}x = \tilde{b} \\
& x \geq 0
\end{aligned}
\tag{6.69}
$$

where c, a, and b are parameters contaminated with epistemic uncertainty, and vector x denotes the continuous variables. Without loss of generality, the imprecise parameters are modeled by trapezoidal fuzzy numbers, specified by four prominent points. For example, $\tilde{a}$ is described as $\left(\xi_{(1)}^a, \xi_{(2)}^a, \xi_{(3)}^a, \xi_{(4)}^a\right)$. Notably, the possibility distribution is extracted based on expert's knowledge, and in the case of $\xi_{(2)} = \xi_{(3)}$ the attributed trapezoidal possibility distribution reduces to a triangular one.

6.5.1.1 Handling uncertainty in the chance constraints

When the LHS or RHS coefficients are subject to uncertainty, such as the first constraint of (6.69), the necessity (Nec) and possibility (Pos) measures can be adopted to cope with the chance constraints:

$$\tilde{a}x \leq b \rightarrow \begin{cases} Nec\left(\tilde{a}x \leq b\right) \geq \beta \\ Poss\left(\tilde{a}x \leq \tilde{b}\right) \geq \beta \end{cases} \tag{6.70}$$

where parameter β is the minimum confidence level of satisfying the corresponding chance constraint, which is determined by DM. The possibility and necessity of fuzzy number $\tilde{\xi}$ are defined based on Dubois (1987) and Dubois and Prade (1988):

$$Pos\left\{\tilde{\xi} \leq r\right\} = \sup_{x \leq r} \mu_{\tilde{\xi}}(x), \tag{6.71}$$

$$Nec\left\{\tilde{\xi} \leq r\right\} = 1 - \sup_{x \succ r} \mu_{\tilde{\xi}}(x) = 1 - Pos\left\{\tilde{\xi} \succ r\right\}. \tag{6.72}$$

Accordingly, $Pos\left\{\tilde{\xi} \leq r\right\}$ and $Nec\left\{\tilde{\xi} \leq r\right\}$ can be stated as follows:

$$Pos\left\{\tilde{\xi} \leq r\right\} = \begin{cases} 1, & \xi_{(2)} \leq r, \\ \dfrac{r - \xi_{(1)}}{\xi_{(2)} - \xi_{(1)}}, & \xi_{(1)} \leq r \leq \xi_{(2)}, \\ 0, & \xi_{(1)} \geq r, \end{cases} \tag{6.73}$$

$$Nec\left\{\tilde{\xi} \leq r\right\} = \begin{cases} 1, & \xi_{(4)} \leq r, \\ \dfrac{r - \xi_{(3)}}{\xi_{(4)} - \xi_{(3)}}, & \xi_{(3)} \leq r \leq \xi_{(4)}, \\ 0, & \xi_{(3)} \geq r. \end{cases} \tag{6.74}$$

It can be shown that if $\beta > 0.5$, the crisp equivalent of the chance constraint based on the necessity (Nec) and possibility (Pos) measures is formulated as follows (Inuiguchi & Ramik, 2000; Liu & Iwamura, 1998):

$$Pos\{\tilde{a}x \leq b\} \geq \beta$$

$$\Leftrightarrow \frac{b - \xi_{(1)}^a}{\xi_{(2)}^a - \xi_{(1)}^a} \geq \beta \qquad (6.75)$$

$$\Leftrightarrow b \geq (1-\beta)\xi_{(1)}^a + \beta\xi_{(2)}^a,$$

$$Nec\{\tilde{a}x \leq b\} \geq \beta$$

$$\Leftrightarrow \frac{b - \xi_{(3)}^a}{\xi_{(4)}^a - \xi_{(3)}^a} \geq \beta \qquad (6.76)$$

$$\Leftrightarrow \beta \geq (1-\beta)\xi_{(3)}^a + \beta\xi_{(4)}^a.$$

It is worth noting that, because of the mandatory nature of satisfying the constraints, applying the necessity measure is more meaningful to cope with chance constraints.

Next, the second and third constraints of (6.69) in which uncertainty appears in both the LHS and RHS are defuzzified. According to the ranking method proposed by Jiménez (1996), for any pair of fuzzy numbers $\tilde{a}$ and $\tilde{b}$, the preference degree of $\tilde{a}$ over $\tilde{b}$ or the degree in which $\tilde{a}$ is greater than $\tilde{b}$ is defined as follows:

$$\mu(\tilde{a},\tilde{b}) = \begin{cases} 0 & if & E_2^a - E_1^b < 0 \\ \dfrac{E_2^a - E_1^b}{E_2^a - E_1^b - \left(E_1^a - E_2^b\right)} & if & 0 \in \left[E_1^a - E_2^b, E_2^a - E_1^b\right] \\ 1 & if & E_1^a - E_2^b > 0 \end{cases} \quad (6.77)$$

where $\left[E_1^a, E_2^a\right]$ and $\left[E_1^b, E_2^b\right]$ are the expected intervals calculated as:

$$\left[E_1^a, E_2^a\right] = \left[\frac{\xi_{(1)}^a + \xi_{(2)}^a}{2}, \frac{\xi_{(3)}^a + \xi_{(4)}^a}{2}\right] \qquad (6.78)$$

$$\left[E_1^b, E_2^b\right] = \left[\frac{\xi_{(1)}^b + \xi_{(2)}^b}{2}, \frac{\xi_{(3)}^b + \xi_{(4)}^b}{2}\right] \qquad (6.79)$$

When $\mu(\tilde{a},\tilde{b}) \geq \alpha$, it is said that $\tilde{a}$ is greater than or equal to $\tilde{b}$ in degree of α and is shown as $\tilde{a} \geq_\alpha \tilde{b}$. With this definition, the inequality constraint in (6.69) is rewritten as:

$$\frac{E_2^{ax} - E_1^{b}}{E_2^{ax} - E_1^{b} - \left(E_1^{ax} - E_2^{b}\right)} \geq \alpha \tag{6.80}$$

Which is equivalent to the following constraint based on Zadeh's minimum extension principle (see Babazadeh, Razmi, Pishvaee, & Omega, 2017):

$$\left[(1-\alpha)E_2^{a} - \alpha E_1^{a}\right]x \geq \alpha E_2^{b} + (1-\alpha)E_2^{b} \tag{6.81}$$

Regarding the equality constraint in (6.69), it is said that $\tilde{a}$ is equal (indifferent) to $\tilde{b}$ in degree of α if the following two relations hold simultaneously:

$$\tilde{a} \geq_{\frac{\alpha}{2}} \tilde{b}, \quad \tilde{a} \leq_{\frac{\alpha}{2}} \tilde{b} \tag{6.82}$$

Which can be rewritten as:

$$\frac{\alpha}{2} \leq \frac{E_2^{ax} - E_1^{b}}{E_2^{ax} - E_1^{b} - \left(E_1^{ax} - E_2^{b}\right)} \leq 1 - \frac{\alpha}{2} \tag{6.83}$$

And according to Zadeh's minimum extension principle, we have:

$$\begin{aligned}
\left[\left(1 - \frac{\alpha}{2}\right)E_2^{a} - \frac{\alpha}{2}E_1^{a}\right]x &\geq \frac{\alpha}{2}E_2^{b} - \left(1 - \frac{\alpha}{2}\right)E_1^{b} \\
\left[\frac{\alpha}{2}E_2^{a} - \left(1 - \frac{\alpha}{2}\right)E_1^{a}\right]x &\leq \left(1 - \frac{\alpha}{2}\right)E_2^{b} - \frac{\alpha}{2}E_1^{b}
\end{aligned} \tag{6.84}$$

6.5.1.2 Handling uncertainty in the objective function

In order to deal with the uncertain coefficients of the objective function, the following three methods can be applied:

(1) Optimizing the possibilistic mean of the objective function as the following problem:

$$\max_{x} z = \tilde{c}x \rightarrow \max_{x} EV[z] \tag{6.85}$$

According to Yager (1981), Dubois and Prade (1987), and Heilpern (1992), it is assumed the fuzzy number $\tilde{c} = \left(\xi_{(1)}^{c}, \xi_{(2)}^{c}, \xi_{(3)}^{c}, \xi_{(4)}^{c}\right)$ is defined by the following general membership function:

$$\mu_{\tilde{\xi}}(x) = \begin{cases} f_{\tilde{\xi}}(x) & if & \xi_{(1)}^c \le x < \xi_{(2)}^c \\ 1 & if & \xi_{(2)}^c \le x < \xi_{(3)}^c \\ g_{\tilde{\xi}}(x) & if & \xi_{(3)}^c \le x < \xi_{(4)}^c \\ 1 & if & x < \xi_{(1)}^c \ or \ x > \xi_{(4)}^c \end{cases} \tag{6.86}$$

Based on the Choquet integral (Dubois & Prade, 1987; Heilpern, 1992), which can be applied to convert a fuzzy number to an interval, the upper (i.e., $E^*(\tilde{c})$) and lower (i.e., $E_*(\tilde{c})$) expected values of fuzzy number $\tilde{\xi}$ are formulated as follows:

$$E^*(\tilde{c}) = \xi_{(3)}^c + \int_{\xi_{(3)}^c}^{\xi_{(4)}^c} g_{\tilde{\xi}}(x)\,dx \tag{6.87}$$

$$E_*(\tilde{c}) = \xi_{(2)}^c - \int_{\xi_{(1)}^c}^{\xi_{(2)}^c} f_{\tilde{\xi}}(x)\,dx \tag{6.88}$$

The interval-valued possibilistic mean of $\tilde{c}$ is then defined as follows (Babazadeh et al., 2017; Heilpern, 1992):

$$EI[\tilde{c}] = \left[E_*(\tilde{\xi}), E^*(\tilde{\xi}) \right] \tag{6.89}$$

The possibilistic mean of $\tilde{c}$ is defined as the middle point of its corresponding interval-valued mean:

$$EV[\tilde{c}] = \frac{E_*(\tilde{\xi}) + E^*(\tilde{\xi})}{2} \tag{6.90}$$

Since $\tilde{c}$ is a trapezoidal fuzzy number, $EV[\tilde{c}]$ is rewritten as follows:

$$EV[\tilde{c}] = \frac{\xi_{(1)}^c + \xi_{(2)}^c + \xi_{(3)}^c + \xi_{(4)}^c}{4} \tag{6.91}$$

(2) Treating the objective function as a chance constraint considering the following optimization problem:

$$\max_x z = \tilde{c}x \rightarrow \begin{array}{c} \max_x u \\ Nec/Pos(\tilde{c}x \ge u) \ge \alpha \end{array} \tag{6.92}$$

where the necessity or possibility measure is used to cope with the chance constraint based on the confidence level of α.

(3) Determining a target value for the objective function and maximizing the confidence level of the associated chance constraint:

$$\max_x \ z = \tilde{c}x \rightarrow \quad \begin{aligned} &\max_x \alpha \\ &Nec/Pos\left(\tilde{c}x \geq \text{target}\right) \geq \alpha \end{aligned} \qquad (6.93)$$

It is worth noting that α is a decision variable in this problem.

6.5.2 Flexible programming model

Flexible programming models can be classified into two main categories: (1) symmetric and (2) nonsymmetric (Naderi et al., 2016). The first category sets a flexible target to maximize/minimize as an objective function rather than strictly optimize the original objective function. The second category assumes there is flexibility in satisfying the constraints of the optimization problem, which leads to soft constraints instead of constraints with sharp boundaries (Pishvaee et al., 2012). In the soft constraints, symbols $\tilde{\leq}, \tilde{=}, \tilde{\geq}$ are introduced to show that a certain amount of deviation is allowed in the constraints. It is worth noting that the fuzzy distributions of flexible objectives and constraints are subjectively determined according to the preferences of the DM.

A following flexible programming problem can be formulated as (Sahinidis, 2004):

$$\begin{aligned} &\widetilde{\min} \quad cx \\ &s.t. \quad Ax \tilde{\geq} d \\ &\quad\quad\ x \geq 0. \end{aligned} \qquad (6.94)$$

Assuming that z is a flexible goal the decision maker would like to reach, problem (6.94) is recast as:

$$\begin{aligned} &Find \quad x \\ &s.t. \quad cx \tilde{\leq} z \\ &\quad\quad\ Ax \tilde{\geq} d \\ &\quad\quad\ x \geq 0. \end{aligned} \qquad (6.95)$$

By introducing the new variable λ that controls the minimum satisfaction level of soft constraints, the above model can be rewritten as follows:

$$
\begin{aligned}
\max \quad & \lambda \\
s.t. \quad & cx \le z_u + \lambda\left(z_l - z_u\right) \\
& Ax \ge d - \tilde{t}\left(1 - \lambda\right) \\
& 0 \le \lambda \le 1 \\
& x \ge 0
\end{aligned}
\tag{6.96}
$$

where z_l and z_u are lower and upper bounds of the flexible objective, respectively, and $\tilde{t}$ is a fuzzy number representing the acceptable violation of the corresponding soft constraint.

If we assume that $\tilde{t}$ is a trapezoidal fuzzy number, which is described with the three prominent points as $\tilde{t} = \left(t^p, t^m, t^o\right)$, it can be defuzzified with the help of the fuzzy ranking method (Yager, 1979; Yager, 1981):

$$
\left(t^m + \frac{\varphi_t - \varphi'_t}{3}\right)
\tag{6.97}
$$

where φ_t and φ'_t are the lateral margins of fuzzy number $\tilde{t}$ that are defined as follows:

$$
\varphi_t = t^o - t^m
\tag{6.98}
$$

$$
\varphi'_t = t^m - t^p.
\tag{6.99}
$$

By using the defuzzified value of $\tilde{t}$, the equivalent crisp model of model (6.94) is obtained as follows:

$$
\begin{aligned}
\max \quad & \lambda \\
s.t. \quad & cx \le z_u + \lambda\left(z_l - z_u\right) \\
& Ax \ge d - \left(t^m + \frac{\varphi_t - \varphi'_t}{3}\right)\left(1 - \lambda\right) \\
& 0 \le \lambda \le 1 \\
& x \ge 0
\end{aligned}
\tag{6.100}
$$

6.5.3 Robust possibilistic programming

According to materials presented in Section 6.5.2, first a basic possibilistic programming problem is developed and then it is extended to a robust possibilistic programming problem.

Let us consider a general form of a supply chain optimization problem:

$$
\begin{aligned}
\min \quad & z = \tilde{f}y + \tilde{c}x \\
s.t. \quad & Ax \geq \tilde{d} \\
& Bx = 0 \\
& Sx \leq \tilde{N}y \\
& Tx \leq 1 \\
& y \in \{0,1\}, x \geq 0
\end{aligned}
\tag{6.101}
$$

where vectors x and y denote the continuous and binary variables, respectively, vectors f and c correspond to fixed costs, variable costs, and demand, respectively, and matrices A, S, B, and N are coefficient matrices of the constraints. Without loss of generality, it is assumed that uncertainty in coefficient matrix N and vectors f, c, and d is modeled as fuzzy numbers.

In order to develop the basic possibilistic chance constrained programming model, the expected value operator is adopted to cope with imprecise parameters in the objective function, where the necessity measure is applied to deal with chance constraints containing uncertain parameters. With these assumptions, the possibilistic programming model is formulated as follows:

$$
\begin{aligned}
\min \quad & E[z] = E\left[\tilde{f}\right]y + E\left[\tilde{c}\right]x \\
s.t. \quad & Nec\left\{Ax \geq \tilde{d}\right\} \geq \alpha \\
& Bx = 0 \\
& Nec\left\{Sx \leq \tilde{N}y\right\} \geq \beta \\
& Tx \leq 1 \\
& y \in \{0,1\}, x \geq 0
\end{aligned}
\tag{6.102}
$$

Using trapezoidal possibility distributions to characterize the imprecise parameters, the equivalent crisp model is obtained as follows (Dubois & Prade, 1987; Heilpern, 1992; Inuiguchi & Ramik, 2000; Liu & Iwamura, 1998):

$$\min \quad E[z] = \left(\frac{f_{(1)} + f_{(2)} + f_{(3)} + f_{(4)}}{4} \right) \gamma + \left(\frac{c_{(1)} + c_{(2)} + c_{(3)} + c_{(4)}}{4} \right) x$$

$$\text{s.t.} \quad Ax \geq (1 - \alpha) d_{(3)} + \alpha d_{(4)}$$

$$Bx = 0 \tag{6.103}$$

$$Sx \leq \left[(1 - \beta) N_{(2)} + \beta N_{(1)} \right] \gamma$$

$$Tx \leq 1$$

$$\gamma \in \{0,1\}, \, x \geq 0$$

where it is assumed that the confidence level of satisfying the chance constraints is greater than 0.5 (i.e., $\alpha, \beta > 0.5$).

The main drawbacks of the basic chance constrained possibilistic programming model can be summarized as follows (Pishvaee et al., 2012):

1. Since the DM should determine the values of α, β (i.e., the chance constraints' confidence levels) in a subjective manner, there is no guarantee that the selected value for each confidence level is optimal.

2. In the related literature, it is proposed that the DM can determine the confidence levels interactively such that at the beginning, the DM determines the initial values for each confidence level, then varies the values and investigates their impact on model outputs in interactive experiments to obtain final values.

3. Increasing the number of chance constraints leads to a significant increase in the required experiments to determine the suitable values of confidence levels. Moreover, determining a confidence level for each chance constraint is not applicable, and the DM usually specifies one confidence level for all related constraints or a category of them.

4. The objective function of the model tries to minimize the expected value and therefore improve the average performance of the system. Indeed, controlling the deviation of the objective function values from its average value in some realizations are not considered in the model, which may result in high risks and costs for the DM.

To address the above problems, robust possibilistic programming has been introduced by Pishvaee et al. (2012). It takes advantage of both robust optimization theory and fuzzy mathematical programming by incorporating the concepts of feasibility robustness and solution robustness into the possibilistic programming. Solution robustness means the value of the objective function remains close to the optimum value for (almost) all possible realizations of imprecise parameters, while feasibility robustness means the solution remains feasible for (almost) all possible realizations of imprecise parameters (Mulvey et al., 1995).

Robust optimization approaches are classified into three categories: (1) hard worst-case robust programming, (2) soft worst-case robust programming, and (3) realistic robust programming. The first approach tries to provide maximum protection against uncertainty while ignoring the possibility of infeasibility. In terms of solution robustness, this approach minimizes the worst possible value of the objective function (i.e., min-max logic), and therefore prevents the violation of the objective function value from the optimal value obtained from solving the related mathematical programming model. The second approach aims to minimize the worst possible value of the objective function. However, satisfying all the constraints in their worst-case value is not enforced in this approach. The last approach tries to make a reasonable trade-off between the solution robustness and feasibility robustness (i.e., cost-benefit logic). The violation of constraints is allowed in the realistic robust programming approach, while the obtained solution is feasible and close to optimal for most of the possible realizations of imprecise parameters. Various versions of robust possibilistic programming models, which cover realistic, hard worst-case, and soft worst-case robust programming approaches are presented in the following sections.

6.5.3.1 Realistic robust possibilistic programming

The realistic robust possibilistic programming version of model (6.102) is formulated as follows:

$$
\begin{aligned}
\min \quad & E[z] + \eta\left(z_{max} - z_{min}\right) + \delta\left[d_{(4)} - (1-\alpha)d_{(3)} - \alpha d_{(4)}\right] \\
& + \pi\left[\beta N_{(1)} + (1-\beta)N_2 - N_{(1)}\right]\gamma
\end{aligned}
$$

$$
s.t. \quad Ax \geq (1-\alpha)d_{(3)} + \alpha d_{(4)}
$$

$$
Bx = 0 \tag{6.104}
$$

$$
Sx \leq \left[\beta N_{(1)} + (1-\beta)N_2\right]\gamma
$$

$$
Tx \leq 1
$$

$$
\gamma \in \{0,1\}, \; x \geq 0, 0.5 < \alpha, \beta \leq 1
$$

The first term of the objective function (i.e., the expected value of z) tries to minimize the average (expected) performance of the system, and the second term indicates the difference between the maximum and minimum possible values of z, which are respectively defined as

$$z_{max} = f_{(4)}y + c_{(4)}x \tag{6.105}$$

$$z_{min} = f_{(1)}y + c_{(1)}x \tag{6.106}$$

Indeed, the second term, weighted by the adjustable parameter η, controls solution robustness by minimizing the maximum deviations of the objective functions from the expected optimal value. The third and fourth terms measure the difference between the values used in the RHS of the chance constraints and their worst possible values, thereby controlling the possible violations in the constraints (i.e., feasibility robustness). The unsatisfied constraints are penalized by the parameters δ and π.

The minimum confidence levels of the chance constraints (i.e., α, β) are variables that are optimized by the model without requiring subjective opinions about the confidence levels. It could be concluded that the proposed model tries to make a reasonable trade-off between three components in the objective function: (1) average performance of the concerned system, (2) solution robustness, and (3) feasibility robustness. Notably, the weight of each component can be regulated based on the DM's preferences.

It can be observed that multiplying decision variables in the last term of the objective function leads to the nonlinearity of the model. However, this model can be converted into an equivalent linear model by defining a nonnegative auxiliary variable $v = \beta y$ and adding the following constraints to the model:

$$\begin{aligned}
&v \leq My, \\
&v \geq M(y-1) + \beta, \\
&v \leq \beta, \\
&v \geq 0,
\end{aligned} \tag{6.107}$$

where M is a predetermined sufficient large number, and these constraints guarantee that if $y = 0$ then v is equal to zero; and if $y = 1$ then v is equal to β. Accordingly, the linear RPP-I model is obtained as follows:

$$\text{min} \quad \begin{aligned} &E[z] + \gamma\left(z_{max} - z_{min}\right) + \delta\left[d_{(4)} - (1-\alpha)d_{(3)} - \alpha d_{(4)}\right] \\ &+ \pi\left[\upsilon N_{(1)} + (\gamma - \upsilon)N_2 - N_{(1)}\gamma\right] \end{aligned}$$

$$\begin{aligned} s.t. \quad & Ax \geq (1-\alpha)d_{(3)} + \alpha d_{(4)} \\ & Bx = 0 \\ & Sx \leq \upsilon N_{(1)} + (\gamma - \upsilon)N_2 \\ & \upsilon \leq M\gamma \\ & \upsilon \geq M(\gamma - 1) + \beta \\ & \upsilon \leq \beta \\ & Tx \leq 1 \\ & \gamma \in \{0,1\}, \ x, \upsilon \geq 0, 0.5 < \alpha, \beta \leq 1 \end{aligned} \quad (6.108)$$

As mentioned above, the role of the second term in the objective function is to minimize the maximum deviation, covering both over and under differences from the expected optimal value. However, it is likely that the DM is not sensitive to both over and under deviations in some real-life problems. For instance, considering total costs objective function, it may be desirable for DM to obtain lower total costs rather than the expected optimal value of the objective function in each realization. Therefore, the objective function can be reformulated to deal with such situations (Pishvaee et al., 2012):

$$\begin{aligned} &\text{min } E[z] + \gamma\left(z_{max} - E[z]\right) + \delta[d_{(4)} - (1-\alpha)d_{(3)} - \alpha d_{(4)}] \\ &+ \pi\left[vN_{(1)} + (\gamma - v)N_{(2)} - N_{(1)}\gamma\right] \end{aligned} \quad (6.109)$$

6.5.3.2 Hard worst-case robust possibilistic programming

The worst-case robust programming approach focuses on immunizing the solution against all possible realizations of the uncertain parameters and therefore ensures maximum feasibility robustness. On the other hand, in terms of optimality robustness, the DM tries to minimize the worst-case value of the objective function among all possible realizations. The hard worst-case robust possibilistic programming model is given as follows:

$$\min \quad \sup(z)$$
$$\text{s.t.} \quad Ax \geq \sup(\tilde{d})$$
$$Bx = 0$$
$$Sx \leq \inf(\tilde{N})\gamma \qquad (6.110)$$
$$Tx \leq 1$$
$$\gamma \in \{0,1\}, \ x \geq 0$$

Assuming the trapezoidal possibility distribution for uncertain parameters, model (6.110) can be rewritten as:

$$\min \quad z_{max}$$
$$\text{s.t.} \quad Ax \geq d_{(4)}$$
$$Bx = 0$$
$$Sx \leq N_{(1)}\gamma \qquad (6.111)$$
$$Tx \leq 1$$
$$\gamma \in \{0,1\}, \ x \geq 0$$

As can be seen, the above model does not depend on the type of possibility distribution of uncertain parameters since it adopts the extreme points of uncertain parameters. Indeed, knowing the support of the concerned possibility distribution is adequate in this model.

6.5.3.3 Soft worst-case robust possibilistic programming

Althogh the hard worst-case robust possibilistic programming model ensures the maximum feasibility robustness (i.e., the highest degree of protection), this protection comes at a significant cost. To make a trade-off between robustness and its cost, the soft worst-case version is developed as follows:

$$\min \quad Z_{max} + \delta[d_{(4)} - (1-\alpha)d_{(3)} - \alpha d_{(4)}] + \pi\left[vN_{(1)} + (\gamma - v)N_{(2)} - N_{(1)}\gamma\right]$$

$$\text{s.t.} \quad Ax \geq (1-\alpha)d_{(3)} + \alpha d_{(4)}$$

$$Bx = 0$$

$$Sx \leq vN_{(1)} + (\gamma - v)N_2$$

$$v \leq M\gamma \qquad\qquad\qquad\qquad (6.112)$$

$$v \geq M(\gamma - 1) + \beta$$

$$v \leq \beta$$

$$Tx \leq 1$$

$$\gamma \in \{0,1\}, \; x, v \geq 0, \; 0.5 < \alpha, \beta \leq 1$$

Similar to the hard worst-case approach, this model seeks to minimize the worst possible value of the objective function but it allows a level of constraint violation that is automatically determined based on the penalty terms in the objective functions, thereby reducing the cost of feasibility robustness.

It is worth mentioning that other versions of robust possibilistic programming such as mixed possibilistic-flexible programming models can be found in Pishvaee and Khalaf (2016).

6.6 Literature review of uncertainty modeling approaches in biofuel supply chain

Based on the proposed typology of biomass supply chain risks in Chapter 4, the risk sources of biomass supply chains can be categorized into three groups: (1) internal risks to the supply chain entities, (2) network-related risks, and (3) external risks to supply chain, which are depicted in Table 6.1. There are various approaches to coping with uncertainty, ranging from simple approaches such as sensitivity analysis to advanced optimization approaches such as stochastic programming, fuzzy programming, robust optimization, and data-driven optimization. It should be noted that sensitivity analysis is classified as an inactive method in managing the uncertainty in parameters because it only studies how the uncertainty in input parameters can impact on the output of a mathematical model while optimization approaches seek to find an optimal solution that performs well in the face of uncertainty.

A summary of the recent literature published on biomass supply chain design and planning under uncertainty is provided in Table 6.2, classifying

Table 6.2 Review of biomass supply chain design and planning studies considering uncertainties.

	Source of uncertainty								Uncertainty modeling approach														Case study
	Internal risks to SC entities				Network-related risks			External risks to SC	SP			FP			Set-based RO				DDO				
Reference	Technology performance	Capital and operating cost	Environmental/social impacts	Yield	Biomass supply	Supply- and demand-side transportation	Biofuel price/demand	Environment/economic/socio-political	Scenario-based SP	Continuous SP	Robust scenario-based	Flexible programming	Possibilistic programming	Robust fuzzy programming	Box uncertainty set	Ellipsoidal uncertainty set	Polyhedral uncertainty set	Adjustable robust optimization	Distributionally robust optimization	Data-driven robust optimization	Sensitivity analysis		
Dal-Mas, Giarola, Zamboni, and Bezzo (2011)					★		★		★														-Corn -Northern Italy
Kim, Realff, and Lee (2011)		★		★	★	★	★		★												★		-Forest biomass -The southeastern part of the United States
An, Wilhelm, and Searcy (2011)				★	★		★														★		-Lignocellulosic biomass -Central Texas

(Continued)

Table 6.2 Review of biomass supply chain design and planning studies considering uncertainties. (*Cont.*)

Reference								Biomass/Location
Chen and Fan (2012)		★		★	★			-Lignocellulosic biomass - California
Gebreslassie, Yao, and You (2012)		★		★	★			-Lignocellulosic biomass -Illinois
Kostin, Guillén-Gosálbez, Mele, Bagajewicz, and Jiménez (2012)		★		★	★			-Sugarcane -Argentina
Walther, Schatka, and Spengler (2012)		★		★	★	★		-Lignocellulosic biomass -Northern Germany
Giarola, Bezzo, and Shah (2013)		★	★	★	★			-Corn, switchgrass, crop residues, -Northern Italy
Sharma, Ingalls, Jones, Huhnke, and Khanchi (2013)		★	★	★				-Switchgrass -Kansas
Osmani and Zhang (2013)		★		★	★	★		-Switchgrass and crop residues -North Dakota
Osmani and Zhang (2014)		★	★	★	★			-Crop residues and woody biomass -Midwestern region of US
Tong, Gong, Yue, and You (2013)				★	★		★	-Crop residues, energy crops, wood residues
Tong, Gleeson, Rong, and You (2014)	★			★	★		★	-Crop residues, energy crops, wood residues -Illinois

(Continued)

Reference	Feedstock / Location
Li and Hu (2014)	–Corn stover –Iowa
Marufuzzaman, Eksioglu, and Huang (2014)	–Wastewater sludge –Mississippi
Azadeh, Arani, and Dashti (2014)	–lignocellulosic biomass
Tong, You, and Rong (2014)	–crop residues, energy crops, wood residues –Illinois
Gonela, Zhang, Osmani, and Onyeaghala (2015b)	–Corn, switchgrass, corn stover –North Dakota
Gonela, Zhang, and Osmani (2015a)	–Corn, switchgrass –North Dakota
Bairamzadeh, Pishvaee, and Saidi-Mehrabad (2015)	–Corn stover, wheat straw –Iran
Santibañez-Aguilar, Morales-Rodriguez, González-Campos, and Ponce-Ortega (2016)	–Wood Chips, sugar cane, Sweet Sorghum, Jatropha, … –Mexico
Shabani and Sowlati (2016)	–Forest-based biomass –Canada
Mohseni, Pishvaee, and Sahebi (2016)	–Microalgae – Iran
Mohseni and Pishvaee (2016)	–Microalgae – Iran

Table 6.2 Review of biomass supply chain design and planning studies considering uncertainties. (*Cont.*)

Reference	Feedstock / Location
Gong, Garcia, and You (2016)	-Soybean, corn, sugar cane, corn stover, hardwood, softwood, switchgrass, microalgae
Poudel, Marufuzzaman, and Bian (2016b)	-Corn stover, forest residues -Mississippi and Alabama
Babazadeh et al. (2017)	-Jatropha -Iran
Hu, Lin, Wang, and Rodriguez (2017)	-Miscanthus -Illinois
Zamar, Gopaluni, Sokhansanj, and Newlands (2017)	-Forest biomass -Numerical example
Quddus, Hossain, Mohammad, Jaradat, and Roni (2017)	-Forest residues, corn stover, miscanthus -Mississippi and Alabama
Bairamzadeh, Saidi-Mehrabad, and Pishvaee (2018b)	-Wheat straw, rice straw, corn stover, barley straw -Iran
Fattahi and Govindan (2018)	-Agricultural residues, forest biomass -Iran
Ahmed and Sarkar (2018)	-Corn stover -Numerical examples
Nguyen and Chen (2018)	-Miscanthus -Numerical studies

Reference	Biomass type / Location
Alizadeh, Ma, Marufuzzaman, and Yu (2019)	-Corn stover, forest residue, municipal Solid Wastes -Mississippi
Poudel, Quddus, Marufuzzaman, and Bian (2019)	-Corn stover, forest residues -Mississippi and Alabama
Razm, Nickel, Saidi-Mehrabad, and Sahebi (2019)	-Forest/agricultural residues, switchgrass -Iran & Armenia
Abasian, Rönnqvist, and Ouhimmou (2019)	-Forest biomass -Canada
Arabi, Yaghoubi, and Tajik (2019)	-Microalgae -Iran
Mohseni and Pishvaee (2020)	-Wastewater sludge -Iran
Saghaei, Ghaderi, and Soleimani (2020)	-Woody biomass -Mississippi
Ahranjani, Ghaderi, Azadeh, and Babazadeh (2020)	-Rice straw, wheat straw, barley straw, corn stover -Iran

Abbreviations: *DDO*, data-driven optimization; *FP*, fuzzy programming; *RO*, robust optimization; *SP*, stochastic programming.

the previous studies in terms of uncertainty source, uncertainty modeling approach, biomass type, and case study region. As observed in Table 6.2, the number of papers addressing uncertainty in the biomass supply chain optimization models has been significantly increased in the recent decade. Fig. 6.4 displays the classification of the uncertainty modeling approaches applied in these studies. It is clear that stochastic programming is the most commonly used approach to tackle the uncertainty in the biomass supply chain optimization problems. However, due to the lack of reliable historical data in most of the real-life problems, forming the probability distribution of the uncertain parameters required in stochastic programming is impossible. On the other hand, uncertainty set-based robust optimization has emerged as an appealing methodology to deal with uncertainty in recent years. In this regard, in the related studies box uncertainty, polyhedral uncertainty, and box–polyhedral uncertainty sets have been broadly adopted to represent the uncertainty of imprecise parameters. Only a small fraction of the studies has applied fuzzy programming approaches, particularly possibilistic programming and flexible programming despite their ability to cope with epistemic uncertainty. Recent years have witnessed a growing number of studies on data–driven optimization under uncertainty, a powerful optimization approach extracting useful information from the underlying

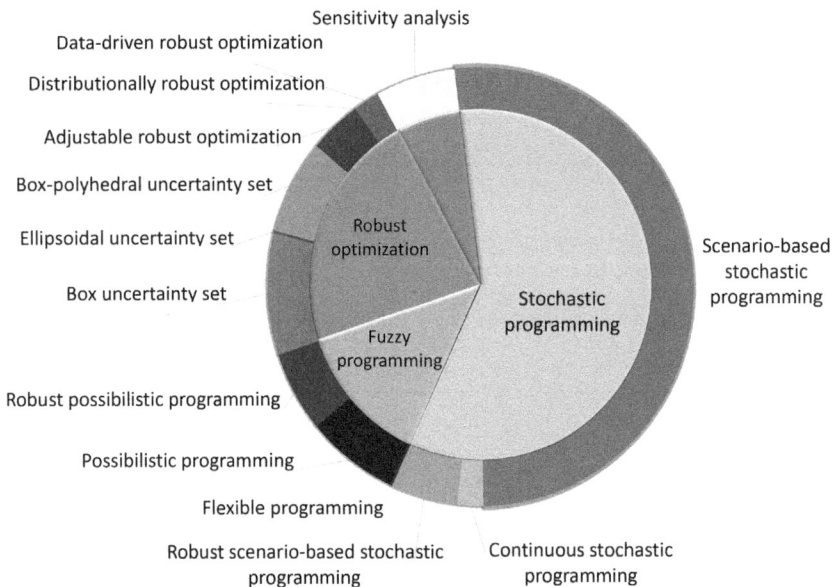

Figure 6.4 *Distribution of uncertainty modeling approaches in the biomass supply chain literature.*

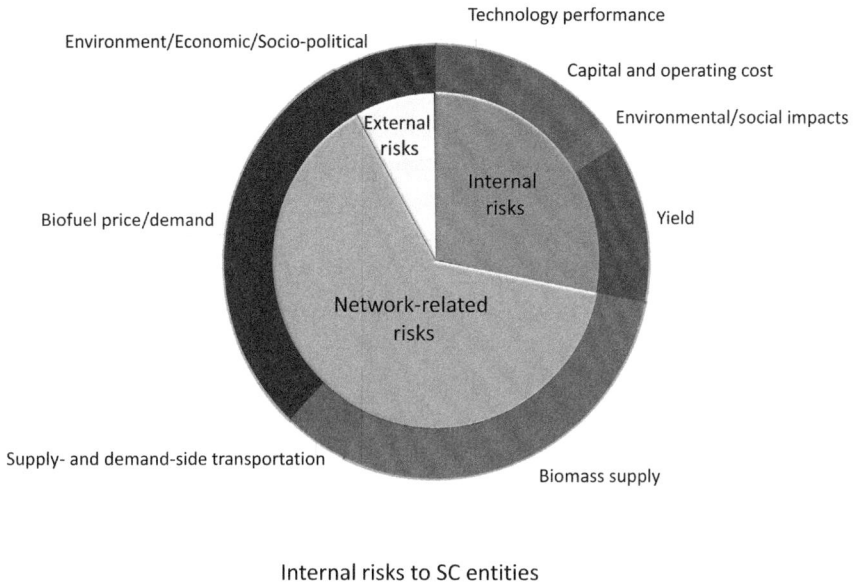

Figure 6.5 *Distribution of the uncertainty types considered in the biomass supply chain literature.*

uncertainty data and incorporating it into optimization. This approach deserves further attention in future work given the development of the biofuel industry and explosion in the availability of data in the upcoming years.

Fig. 6.5 shows the distribution of uncertainty sources that have been analyzed in the biomass supply chain literature. Based on this figure, most of the researches addressed network-related risks in the biomass supply chain optimization models while external risks involving environment, economic, and socio-political uncertainty have the smallest portion in the literature. On the other hand, approximately 60% of the research studies considered uncertainty in biomass supply and price as well as biofuel demand and price. Moreover, it can be observed that only 13% of the studies assumed environmental and social impacts, supply- and demand-side transportation, or technology performance to be uncertain in the biomass supply chain modeling and optimization problems.

6.7 Conclusions

Uncertainty is considered as one of the most crucial factors influencing the performance of the supply chain and, if ignored, can lead to suboptimal performance or even infeasible supply chain configurations. Based on their scope, controllability, predictability, and propagation properties, the risk sources of biomass supply chain are categorized into three groups, including (1) internal risks to the supply chain entities, (2) network-related risks, and (3) external risks to the supply chain. There are three types of uncertainty, namely, randomness uncertainty, epistemic uncertainty, and deep uncertainty, specifically in the context of supply chain optimization. After determining the type of uncertainty, an appropriate optimization approach for coping with uncertainty must be selected. In choosing the most suitable approach for handling uncertainty in the supply chain, the amount of available data plays an important role in determining the type of uncertainty and its corresponding treatment approaches. This chapter provides a comprehensive overview of leading optimization approaches for hedging against various types of uncertainty in the biomass supply chain design and planning models. In this regard, a detailed description of the uncertainty modeling approaches, including stochastic programming, fuzzy programming, uncertainty set-based robust optimization, and data-driven optimization, is presented.

Additionally, the studies utilizing optimization approaches to deal with various types of uncertainty in the biomass supply chain design and planning problems are reviewed and classified in terms of the sources of uncertainty, uncertainty modeling approach, biomass type, and case study region. The findings of this review show that stochastic programming is the most commonly used approach to tackle the uncertain parameters in the biomass supply chain optimization problems. On the other hand, set-based robust optimization approaches, particularly data-driven robust optimization approaches, have emerged as an appealing methodology to deal with uncertainty in recent years while box uncertainty, polyhedral uncertainty, and box-polyhedral uncertainty sets have been broadly adopted to represent the uncertainty of imprecise parameters.

References

Abasian, F., Rönnqvist, M., & Ouhimmou, M. (2019). Forest bioenergy network design under market uncertainty. *Energy*, *188*, 116038.

Ahmed, W., & Sarkar, B. (2018). Impact of carbon emissions in a sustainable supply chain management for a second generation biofuel. *Journal of Cleaner Production*, *186*, 807–820.

Ahranjani, P. M., Ghaderi, S. F., Azadeh, A., & Babazadeh, R. (2020). Robust design of a sustainable and resilient bioethanol supply chain under operational and disruption risks. *Clean Technologies and Environmental Policy*, *22*, 119–151.

Alizadeh, M., Ma, J., Marufuzzaman, M., & Yu, F. (2019). Sustainable olefin supply chain network design under seasonal feedstock supplies and uncertain carbon tax rate. *Journal of Cleaner Production*, *222*, 280–299.

An, H., Wilhelm, W. E., & Searcy, S. W. (2011). A mathematical model to design a lignocel-lulosic biofuel supply chain system with a case study based on a region in Central Texas. *Bioresource Technology, 102*, 7860–7870.

Arabi, M., Yaghoubi, S., & Tajik, J. (2019). A mathematical model for microalgae-based biobutanol supply chain network design under harvesting and drying uncertainties. *Energy, 179*, 1004–1016.

Azadeh, A., Arani, H. V., & Dashti, H. (2014). A stochastic programming approach towards optimization of biofuel supply chain. *Energy, 76*, 513–525.

Babazadeh, R., Razmi, J., Pishvaee, M. S., & Omega, M. R. (2017). *A sustainable second-generation biodiesel supply chain network design problem under risk.* Elsevier.

Bairamzadeh, S., Pishvaee, M. S., & Saidi-Mehrabad, M. (2015). Multiobjective robust pos-sibilistic programming approach to sustainable bioethanol supply chain design under multiple uncertainties. *Industrial & Engineering Chemistry Research, 55*, 237–256.

Bairamzadeh, S., Saidi-Mehrabad, M., & Pishvaee, M. S. (2018a). Modelling different types of uncertainty in biofuel supply network design and planning: A robust optimization approach. *Renewable Energy, 116*, 500–517.

Bairamzadeh, S., Saidi-Mehrabad, M., & Pishvaee, M. S. (2018b). Modelling different types of uncertainty in biofuel supply network design and planning: A robust optimization approach. *Renewable Energy, 116*, 500–517.

Ben-Hur, A., Horn, D., Siegelmann, H. T., & Vapnik, V. (2001). Support vector clustering. *Journal of Machine Learning Research, 2*, 125–137.

Ben-Tal, A., Goryashko, A., Guslitzer, E., & Nemirovski, A. (2004). Adjustable robust solu-tions of uncertain linear programs. *Mathematical Programming, 99*, 351–376.

Ben-Tal, A., & Nemirovski, A. (1999). Robust solutions of uncertain linear programs. *Operations Research Letters, 25*, 1–13.

Bertsimas, D., & Sim, M. (2004). The price of robustness. *Operations Research, 52*, 35–53.

Bertsimas, D., Sim, M., & Zhang, M. (2017). *A practically efficient approach for solving adaptive distributionally robust linear optimization problems.* Manag. Sci.

Bertsimas, D., & Thiele, A. (2006). Robust and data-driven optimization: Modern decision making under uncertainty. In *Models, methods, and applications for innovative decision making.* INFORMS.

Birge, J. R., & Louveaux, F. (2011). *Introduction to stochastic programming.* Springer Science & Business Media.

Chakraborty, D. (2002). Redefining chance-constrained programming in fuzzy environment. *Fuzzy Sets and Systems, 125*, 327–333.

Charnes, A., & Cooper, W. W. (1959). Chance-constrained programming. *Management Science, 6*, 73–79.

Chen, C. -W., & Fan, Y. (2012). Bioethanol supply chain system planning under supply and demand uncertainties. *Transportation Research Part E: Logistics and Transportation Review, 48*, 150–164.

Christopher, M., & Peck, H. (2004). Building the resilient supply chain. *The International Journal of Logistics Management, 15*, 1–14.

Dal-Mas, M., Giarola, S., Zamboni, A., & Bezzo, F. (2011). Strategic design and investment capacity planning of the ethanol supply chain under price uncertainty. *Biomass and Bio-energy, 35*, 2059–2071.

Delage, E., & Ye, Y. (2010). Distributionally robust optimization under moment uncertainty with application to data-driven problems. *Operations Research, 58*, 595–612.

Dubois, D. (1987). Linear programming with fuzzy data. *Analysis of Fuzzy Information, 3*, 241–263.

Dubois, D., & Prade, H. (1987). The mean value of a fuzzy number. *Fuzzy Sets and Systems, 24*, 279–300.

Dubois, D., & Prade, H. (1988). *Possibility theory.* New York: Plenum Press.

Fattahi, M., & Govindan, K. (2018). A multi-stage stochastic program for the sustainable design of biofuel supply chain networks under biomass supply uncertainty and disruption risk: A real-life case study. *Transportation Research Part E: Logistics and Transportation Review*, *118*, 534–567.

Gebreslassie, B. H., Yao, Y., & You, F. (2012). Design under uncertainty of hydrocarbon biorefinery supply chains: Multiobjective stochastic programming models, decomposition algorithm, and a comparison between CVaR and downside risk. *AIChE Journal*, *58*, 2155–2179.

Giarola, S., Bezzo, F., & Shah, N. (2013). A risk management approach to the economic and environmental strategic design of ethanol supply chains. *Biomass and Bioenergy*, *58*, 31–51.

Gonela, V., Zhang, J., & Osmani, A. (2015a). Stochastic optimization of sustainable industrial symbiosis based hybrid generation bioethanol supply chains. *Computers & Industrial Engineering*, *87*, 40–65.

Gonela, V., Zhang, J., Osmani, A., & Onyeaghala, R. (2015b). Stochastic optimization of sustainable hybrid generation bioethanol supply chains. *Transportation Research Part E: Logistics and Transportation Review*, *77*, 1–28.

Gong, J., Garcia, D. J., & You, F. (2016). Unraveling optimal biomass processing routes from bioconversion product and process networks under uncertainty: An adaptive robust optimization approach. *ACS Sustainable Chemistry & Engineering*, *4*, 3160–3173.

Govindan, K., Fattahi, M., & Keyvanshokooh, E. (2017). Supply chain network design under uncertainty: A comprehensive review and future research directions. *European Journal of Operational Research*, *263*, 108–141.

Heilpern, S. (1992). The expected value of a fuzzy number. *Fuzzy Sets and Systems*, *47*, 81–86.

Hester, P. (2012). Epistemic uncertainty analysis: An approach using expert judgment and evidential credibility. *International Journal of Quality, Statistics, and Reliability*, 2012.

Hu, H., Lin, T., Wang, S., & Rodriguez, L. F. (2017). A cyberGIS approach to uncertainty and sensitivity analysis in biomass supply chain optimization. *Applied Energy*, *203*, 26–40.

Inuiguchi, M., & Ramik, J. (2000). Possibilistic linear programming: A brief review of fuzzy mathematical programming and a comparison with stochastic programming in portfolio selection problem. *Fuzzy Sets and Systems*, *111*, 3–28.

Jiménez, M. (1996). Ranking fuzzy numbers through the comparison of its expected intervals. *International Journal of Uncertainty, Fuzziness and Knowledge-Based Systems*, *4*, 379–388.

Kim, J., Realff, M. J., & Lee, J. H. (2011). Optimal design and global sensitivity analysis of biomass supply chain networks for biofuels under uncertainty. *Computers & Chemical Engineering*, *35*, 1738–1751.

Klibi, W., Martel, A., & Guitouni, A. (2010). The design of robust value-creating supply chain networks: A critical review. *European Journal of Operational Research*, *203*, 283–293.

Kostin, A., Guillén-Gosálbez, G., Mele, F., Bagajewicz, M., & Jiménez, L. (2012). Design and planning of infrastructures for bioethanol and sugar production under demand uncertainty. *Chemical Engineering Research and Design*, *90*, 359–376.

Langholtz, M., Webb, E., Preston, B. L., Turhollow, A., Breuer, N., Eaton, L., King, A. W., Sokhansanj, S., Nair, S. S., & Downing, M. (2014). Climate risk management for the US cellulosic biofuels supply chain. *Climate Risk Management*, *3*, 96–115.

Leung, S. C., Tsang, S. O., Ng, W. -L., & Wu, Y. (2007). A robust optimization model for multi-site production planning problem in an uncertain environment. *European Journal of Operational Research*, *181*, 224–238.

Li, Q., & Hu, G. (2014). Supply chain design under uncertainty for advanced biofuel production based on bio-oil gasification. *Energy*, *74*, 576–584.

Li, Z., Ding, R., & Floudas, C. A. (2011). A comparative theoretical and computational study on robust counterpart optimization: I. Robust linear optimization and robust mixed integer linear optimization. *Industrial & Engineering Chemistry Research*, *50*, 10567–10603.

Liu, B., & Iwamura, K. (1998). Chance constrained programming with fuzzy parameters. *Fuzzy Sets and Systems*, *94*, 227–237.

Marufuzzaman, M., Eksioglu, S. D., & Huang, Y. (2014). Two-stage stochastic programming supply chain model for biodiesel production via wastewater treatment. *Computers & Operations Research*, *49*, 1–17.

Mohseni, S., & Pishvaee, M. S. (2016). A robust programming approach towards design and optimization of microalgae-based biofuel supply chain. *Computers & Industrial Engineering*, *100*, 58–71.

Mohseni, S., & Pishvaee, M. S. (2019). Data-driven robust optimization for wastewater sludge-to-biodiesel supply chain design. *Computers & Industrial Engineering*, 105944.

Mohseni, S., & Pishvaee, M. S. (2020). Data-driven robust optimization for wastewater sludge-to-biodiesel supply chain design. *Computers & Industrial Engineering*, *139*, 105944.

Mohseni, S., Pishvaee, M. S., & Sahebi, H. (2016). Robust design and planning of microalgae biomass-to-biodiesel supply chain: A case study in Iran. *Energy*, *111*, 736–755.

Mulvey, J. M., Vanderbei, R. J., & Zenios, S. A. (1995). Robust optimization of large-scale systems. *Operations Research*, *43*, 264–281.

Naderi, M. J., Pishvaee, M. S., & Torabi, S. A. (2016). *Applications of fuzzy mathematical programming approaches in supply chain planning problems. Fuzzy logic in its 50th year*. Springer.

Nemirovski, A., & Shapiro, A. (2007). Convex approximations of chance constrained programs. *SIAM Journal on Optimization*, *17*, 969–996.

Nguyen, D. H., & Chen, H. (2018). Supplier selection and operation planning in biomass supply chains with supply uncertainty. *Computers & Chemical Engineering*, *118*, 103–117.

Ning, C., & You, F. (2017). Data-driven adaptive nested robust optimization: General modeling framework and efficient computational algorithm for decision making under uncertainty. *AIChE Journal*, *63*, 3790–3817.

Ning, C., & You, F. (2019). Optimization under uncertainty in the era of big data and deep learning: When machine learning meets mathematical programming. *Computers & Chemical Engineering*, *125*, 434–448.

Osmani, A., & Zhang, J. (2013). Stochastic optimization of a multi-feedstock lignocellulosic-based bioethanol supply chain under multiple uncertainties. *Energy*, *59*, 157–172.

Osmani, A., & Zhang, J. (2014). Economic and environmental optimization of a large scale sustainable dual feedstock lignocellulosic-based bioethanol supply chain in a stochastic environment. *Applied Energy*, *114*, 572–587.

Pishvaee, M. S., & Khalaf, M. F. (2016). Novel robust fuzzy mathematical programming methods. *Applied Mathematical Modelling*, *40*, 407–418.

Pishvaee, M. S., Rabbani, M., & Torabi, S. A. (2011). A robust optimization approach to closed-loop supply chain network design under uncertainty. *Applied Mathematical Modelling*, *35*, 637–649.

Pishvaee, M. S., Razmi, J., & Torabi, S. A. (2012). Robust possibilistic programming for socially responsible supply chain network design: A new approach. *Fuzzy Sets and Systems*, *206*, 1–20.

Poudel, S. R., Marufuzzaman, M., & Bian, L. (2016a). Designing a reliable bio-fuel supply chain network considering link failure probabilities. *Computers & Industrial Engineering*, *91*, 85–99.

Poudel, S. R., Marufuzzaman, M., & Bian, L. (2016b). A hybrid decomposition algorithm for designing a multi-modal transportation network under biomass supply uncertainty. *Transportation Research Part E: Logistics and Transportation Review*, *94*, 1–25.

Poudel, S. R., Quddus, M. A., Marufuzzaman, M., & Bian, L. (2019). Managing congestion in a multi-modal transportation network under biomass supply uncertainty. *Annals of Operations Research*, *273*, 739–781.

Quddus, M. A., Hossain, N. U. I., Mohammad, M., Jaradat, R. M., & Roni, M. S. (2017). Sustainable network design for multi-purpose pellet processing depots under biomass supply uncertainty. *Computers & Industrial Engineering*, *110*, 462–483.

Rahimian, H., & Mehrotra, S. (2019). Distributionally robust optimization: A review. *arXiv preprint arXiv*. 1908.05659.

Razm, S., Nickel, S., Saidi-Mehrabad, M., & Sahebi, H. (2019). A global bioenergy supply network redesign through integrating transfer pricing under uncertain condition. *Journal of Cleaner Production, 208*, 1081–1095.

Saghaei, M., Ghaderi, H., & Soleimani, H. (2020). Design and optimization of biomass electricity supply chain with uncertainty in material quality, availability and market demand. *Energy*, 117165.

Sahinidis, N. V. (2004). Optimization under uncertainty: State-of-the-art and opportunities. *Computers & Chemical Engineering, 28*, 971–983.

Santibañez-Aguilar, J. E., Morales-Rodriguez, R., González-Campos, J. B., & Ponce-Ortega, J. M. (2016). Stochastic design of biorefinery supply chains considering economic and environmental objectives. *Journal of Cleaner Production, 136*, 224–245.

Shabani, N., & Sowlati, T. (2016). A hybrid multi-stage stochastic programming-robust optimization model for maximizing the supply chain of a forest-based biomass power plant considering uncertainties. *Journal of Cleaner Production, 112*, 3285–3293.

Shang, C., Huang, X., & You, F. (2017). Data-driven robust optimization based on kernel learning. *Computers & Chemical Engineering, 106*, 464–479.

Shang, C., & You, F. (2018). Distributionally robust optimization for planning and scheduling under uncertainty. *Computers & Chemical Engineering, 110*, 53–68.

Sharma, B., Ingalls, R. G., Jones, C. L., Huhnke, R. L., & Khanchi, A. (2013). Scenario optimization modeling approach for design and management of biomass-to-biorefinery supply chain system. *Bioresource Technology, 150*, 163–171.

Soyster, A. L. (1973). Convex programming with set-inclusive constraints and applications to inexact linear programming. *Operations Research, 21*, 1154–1157.

Tong, K., Gleeson, M. J., Rong, G., & You, F. (2014a). Optimal design of advanced drop-in hydrocarbon biofuel supply chain integrating with existing petroleum refineries under uncertainty. *Biomass and Bioenergy, 60*, 108–120.

Tong, K., Gong, J., Yue, D., & You, F. (2013). Stochastic programming approach to optimal design and operations of integrated hydrocarbon biofuel and petroleum supply chains. *ACS Sustainable Chemistry & Engineering, 2*, 49–61.

Tong, K., You, F., & Rong, G. (2014b). Robust design and operations of hydrocarbon biofuel supply chain integrating with existing petroleum refineries considering unit cost objective. *Computers & Chemical Engineering, 68*, 128–139.

Van Der Vaart, J., De Vries, J., & Wijngaard, J. (1996). Complexity and uncertainty of materials procurement in assembly situations. *International Journal of Production Economics, 46*, 137–152.

Walther, G., Schatka, A., & Spengler, T. S. (2012). Design of regional production networks for second generation synthetic bio-fuel-A case study in Northern Germany. *European Journal of Operational Research, 218*, 280–292.

Wiesemann, W., Kuhn, D., & Sim, M. (2014). Distributionally robust convex optimization. *Operations Research, 62*, 1358–1376.

Yager, R. R. (1979). Ranking fuzzy subsets over the unit interval. In *1978 IEEE conference on decision and control including the 17th symposium on adaptive processes, IEEE* (pp. 1435–1437).

Yager, R. R. (1981). A procedure for ordering fuzzy subsets of the unit interval. *Information Sciences, 24*, 143–161.

Yanikoğlu, İ., Gorissen, B. L., & Den Hertog, D. (2019). A survey of adjustable robust optimization. *European Journal of Operational Research, 277*, 799–813.

Yue, D., You, F., & Snyder, S. W. (2014). Biomass-to-bioenergy and biofuel supply chain optimization: Overview, key issues and challenges. *Computers & Chemical Engineering, 66*, 36–56.

Yue, J., Chen, B., & Wang, M. -C. (2006). Expected value of distribution information for the newsvendor problem. *Operations Research, 54*, 1128–1136.

Zamar, D. S., Gopaluni, B., Sokhansanj, S., & Newlands, N. K. (2017). A quantile-based scenario analysis approach to biomass supply chain optimization under uncertainty. *Computers & Chemical Engineering, 97*, 114–123.

Zhang, Y., Jin, X., Feng, Y., & Rong, G. (2018). Data-driven robust optimization under correlated uncertainty: A case study of production scheduling in ethylene plant. *Computers & Chemical Engineering, 109*, 48–67.

CHAPTER 7

Strategic planning in biofuel supply chain under uncertainty

7.1 Introduction

Depending on the time horizon, supply chain planning levels and corresponding decisions can be classified as strategic (long-term), tactical (medium-term), and operational (short-term) (Melo, Nickel, & Saldanha-Da-Gama, 2009). The strategic decisions are those decisions with a long-lasting impact on the supply chain, which include decisions related to the location, capacity, and type of supply chain facilities. In biomass to biofuel supply chains, these decisions involve, for instance, (1) selection of biomass types, (2) finding suitable locations for biomass cultivation, (3) determining the location, number, technology, and capacity of biorefineries, and (4) identifying appropriate insertion points for combining the supply chain of biofuels with that of their petroleum-derived counterparts (Yue, You, & Snyder, 2014). The medium–term planning level in biomass supply chains deals with tactical decisions that are usually made over a time horizon ranging from a few months to one year, such as inventory planning of biomass and biofuel. Operational level decisions are those decisions made within a short-term planning horizon, such as scheduling of harvesting and collection operations. As shown in Fig. 7.1, these three planning levels are not isolated but they are closely connected with each other. Generally speaking, the highest level provides the optimal configuration of the supply chain, serving as a building within which the lower layers determine tactical and operational decisions.

   An overview of strategic decisions in biomass supply chains is provided in Fig. 7.2. The network design of biomass supply chains, as the main strategic decisions, aims to systematically design the entire chain including biomass production and preprocessing, biomass to biofuel conversion, biofuel blending and distribution, and biofuel sales. The first step in a network design project is the identification of potentially suitable locations for the establishment of supply chain facilities such as biorefineries (Melo et al., 2009). The selection of biofuel production capacity and biomass-to-biofuel conversion technology are two important strategic decisions associated with

Biomass to Biofuel Supply Chain Design and Planning under Uncertainty
http://dx.doi.org/10.1016/B978-0-12-820640-9.00007-6

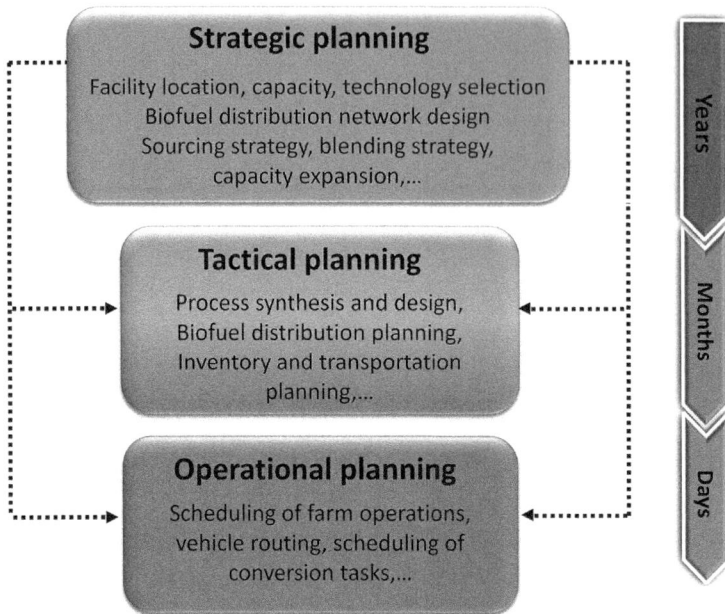

Figure 7.1 *Strategic, tactical, and operational planning levels in biofuel supply chains.*

biorefineries. The importance of capacity determination stems from the fact that, in biofuel supply chains, there is a strong trade-off between biorefinery costs and transportation costs. The reason is that the cost-saving resulted from increasing the capacity of biorefinery due to the economy of scale can be offset by the increased costs of transporting low-density biomass for longer distances from various supply areas to a large-scale biorefinery (You & Wang, 2011). The selection of conversion technology plays a significant role in the economic and environmental performance of biofuel supply chains because different technologies may vary in terms of fixed and operational costs, conversion yield, feedstock requirements, byproducts types, government incentives, and environmental impacts (Bairamzadeh, Saidi-Mehrabad, & Pishvaee, 2018; Marvin, Schmidt, Benjaafar, Tiffany, & Daoutidis, 2012).

The design of the biofuel distribution network, as another strategic decision, consists of determining the number and capacity of storage facilities, blending sites, and distribution centers, as well as the optimal flow of biofuels from biorefineries to the final market. In this regard, the selection of the optimal blending option is considered as one of the important strategic decisions in determining the physical structure of the distribution network in the biomass to biofuel supply chains. In the conventional biofuel supply

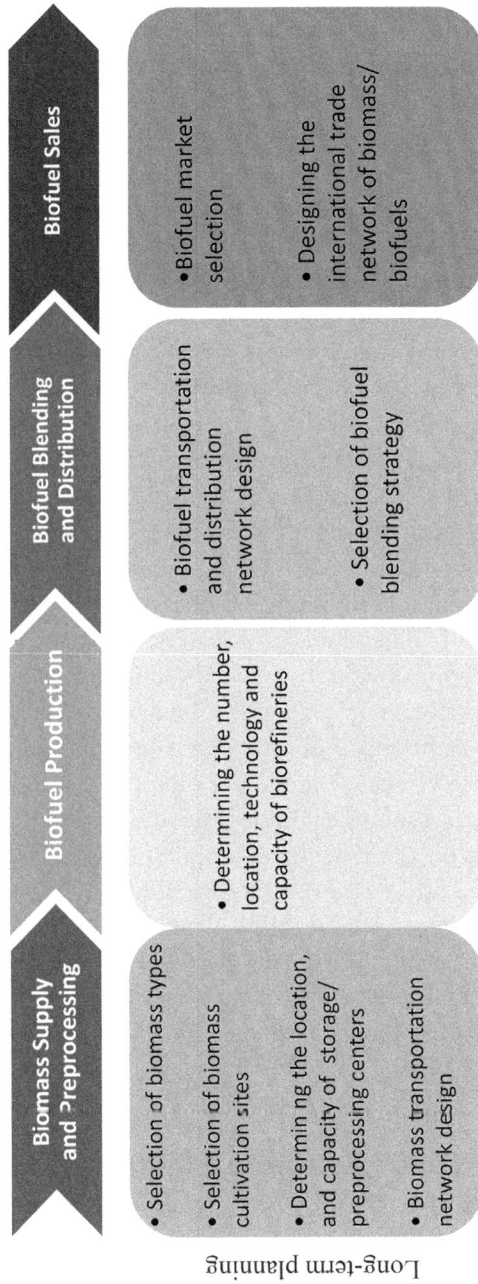

Figure 7.2 *An overview of strategic decisions in biofuel supply chains.*

Long-term planning

Biomass Supply and Preprocessing

- Selection of biomass types
- Selection of biomass cultivation sites
- Determining the location, and capacity of storage/preprocessing centers
- Biomass transportation network design

Biofuel Production

- Determining the number, location, technology and capacity of biorefineries

Biofuel Blending and Distribution

- Biofuel transportation and distribution network design
- Selection of biofuel blending strategy

Biofuel Sales

- Biofuel market selection
- Designing the international trade network of biomass/biofuels

chains, there are two options for blending biofuels with their petroleum fuel counterparts. In the first blending option, biofuel is transported directly from biorefineries to distribution terminals and is blended in different proportions with fuels coming from petroleum refineries. Most previous researches have addressed this option in the design of biomass supply chain structures. In the second option, biofuel is transported from biorefineries to petroleum refineries to be blended with conventional fossil fuels there and then be transported to final costumers through existing pipelines and available distribution channels. The blending options of conventional and advanced hydrocarbon biofuels are discussed in detail in Chapter 3.

The design of biomass supply systems as an important strategic decision determines how to provide the required biomass resources for biorefineries at least cost and in appropriate quality. The selection of biomass type is another strategic decision that cannot be altered in the short term after it is made. This is because energy crops typically require several years to establish. Selecting the type of biomass influences not only the economic feasibility of biofuel production but also brings the consequences for society and the environment in terms of the unemployment rate, welfare, carbon dioxide sequestration, soil erosion, water quality, and others (Cobuloglu & Büyüktahtakin, 2014). Therefore, in choosing the most suitable biomass type, one must take into account various, possibly conflicting, factors such as economic, environmental, and social outcomes of biomass production. Another aspect to be considered is the characteristics and infrastructure of the region or country where the biofuel industry is to be developed because suitable biomass for biofuel production may vary from one location to another depending on the potential availability of biomass resources, geographical conditions, and technological capabilities. Moreover, since relying on a single biomass type increases the vulnerability of the supply chain to biomass supply disruptions, portfolio selection methods can be applied to build a diverse portfolio of biomass sources with the purpose of improving the reliability and efficiency of the feedstock supply system (Kashanian, Pishvaee, & Sahebi, 2020).

7.2 An overview of uncertainties related to strategic decisions

Biomass supply chains are subject to various sources of uncertainty in the upstream, midstream, and downstream of the supply chain, which were described in depth in Chapter 4. Biomass supply chain uncertainties can be categorized into strategic (long-term) and operational (short-term) uncertainties based on the time horizon of impact (Yue & You, 2016). Strategic uncertainties are those uncertainties with long-lasting impacts and low occurrence frequency that remain unchanged for a significant period of time once realized. Examples of long-term uncertainties include changes in government incentives and policies, technology evolution, and disruptions (due to catastrophic events such as hurricane, drought, and earthquake). On the other hand, operational uncertainties are those uncertainties that change more frequently and often need immediate adjustment within short time intervals. Examples of short-term uncertainties involve variations in biomass quality and price, biofuel demand and price, and biofuel production costs (Yue & You, 2016).

Due to the long-term nature of strategic decisions such as the location and capacity of biorefineries, the decision-maker must consider strategic long-term uncertainties at the design phase to immunize strategic decisions against these uncertainties. Moreover, there is a need to hedge against operational uncertainties to enhance the robustness of the supply chain design. Although the impact of operational uncertainties may be low at its time of realization, their cumulative effect over the entire planning horizon could be not trivial, influencing strategic long-term decisions. For instance, in designing the biomass supply chains, it seems that ignoring the variations of biomass yields, which are usually realized in seasonal time periods, does not have a substantial impact on strategic decisions. However, the total deviation between the estimated and realized values of yields parameter within a multiyear time period, which is a typical time horizon for biofuel projects, may affect the required land area for cultivating biomass sources or the capacity of conversion facilities. Short-term uncertainties will be further discussed in Chapters 8 and 9.

A biofuel supply chain design model that copes with various strategic and operational uncertainties, although appealing, is often computationally intractable. Consequently, the uncertain parameters with the most impact on the performance of the supply chain are considered when modeling uncertainty in the supply chain. One of the main uncertainties that must be addressed by biofuel supply chain design models is technology uncertainty, arising from the fact that the biofuel industry is still in its infancy and large-scale production of biofuels has not yet achieved in a commercial scale. As an important strategic uncertainty, technology performance uncertainty stems from concerns that there is no guarantee that biofuel technologies operate and function as efficiently and effectively as expected and refers to issues such as unproven technology, engineering failure, and poor technology implementation (Yatim, Lin, Lam, & Choy, 2017). The importance of this type of risk has been increasingly highlighted in recent years because the biofuel industry is moving from first-generation biofuel technology to second- and third-generation technologies, which are still in development and not matured. Therefore, determining whether to employ the mature conventional technologies with less improving potential or emerging technologies with promising efficiency is one of the most important strategic decisions in the design phase of biomass supply chains, which can be described through discrete scenarios (Yue et al., 2014). In this regard, scenario-based stochastic programming, as a conventional approach to tackle the randomness uncertainty, can be applied to address this issue. Supply chain optimization problems are generally formulated as two-stage stochastic programming models where first-stage decisions corresponding to strategic decisions such as the selection of production technologies must be made before uncertainty is realized and second-stage decisions corresponding to tactical/operational decisions such as the production and transportation of biomass and biofuels can be made after the values of the uncertain parameters are disclosed. Besides, scenario-based robust optimization, which uses several discrete scenarios for describing randomness uncertainty, can also be adopted. It is worth noting that, according to the presented literature review in Chapter 6, a limited number of researches have addressed technology uncertainty in the modeling and optimization of biomass supply chains, including Tong, Gong, Yue, and You (2013), Li and Hu (2014), Marufuzzaman, Eksioglu, and Huang (2014), Hu, Lin, Wang, and Rodriguez (2017), and Nodooshan et al. (2018).

On the other hand, the immaturity of processing technologies and lack of commercial-scale biofuel production have led to high variability in the capital and operating costs associated with biorefineries. For instance, a harmonized cost comparison study examining differences in the reported production costs of microalgae-based biofuel reveals that the capital costs of an integrated biorefinery with an annual capacity of 50 million gallons range from $675.9 to $1332 million, based on a normalized set of input parameters and assumptions (Sun et al., 2011). In addition to the continuous stochastic programming that requires extensive historical data to form probability distributions, fuzzy mathematical programming and uncertainty set based robust optimization, which respectively need the subjective knowledge and variation range of uncertain parameters, can be utilized to handle the cost uncertainties.

Another important strategic uncertainty influencing the performance of biomass supply chains is related to changes in government policies and regulations. Many governments around the world have promulgated supporting policies such as tax exemptions, subsidies, and compulsory consumption schemes to promote renewable energy development, in general, and biofuels, in particular, which can facilitate the transition toward a more sustainable and secure energy future (Pérez, Camargo, Rincón, & Marchant, 2017). However, these policies are likely to be changed or even revoked after the establishment of biomass supply chains due to economic and environmental pressures. Changes or ambiguities in the regulations or policies related to the production of biofuel result in a discrepancy between the realized and predicted returns, which reduces the attractiveness of the biofuel industry to investors and other financing sources. To address this type of uncertainty, scenario-based stochastic programming, and scenario-based robust optimization are appropriate approaches that can be applied to optimize the expected profit over the possible scenarios.

In addition to the risks mentioned above, disruption risks originating from man-made or natural disasters (e.g., terrorist attacks, earthquakes, droughts, and floods) threaten the structure and operation of the supply chain, which can be considered as another type of strategic uncertainties (Marufuzzaman & Ekşioğlu, 2016). A review of studies on supply chain optimization under disruption risks shows that there are two main ways to make the supply chain more resilient against disruptions. The first way is using an optimization approach that often models disruptions as a set of discrete scenarios representing possible changes in supply chain components (e.g., suppliers, transporting links, and production facilities) when a disruption occurs and then determines optimal supply chain decisions under the defined scenarios. The second utilizes resilience strategies with the purpose of reducing supply chain vulnerability. There are various resilience strategies to hedge against disruption risks and enhance the resiliency of biomass supply chains, which can be broadly grouped into three classes. These strategies are described in detail in Chapter 4.

7.3 Identification of candidate locations for supply chain design models

Determining the location of different components of biomass supply chains such as cultivation sites and biorefineries is considered as one of the main strategic decisions and is included in almost all biomass supply chain design models. A common issue that arises during the implementation of these models is how to identify candidate locations for supply chain entities from which the optimal locations are to be determined. The difficulty of this task stems from the fact that the biomass supply chain is often geographically dispersed, and various climatic, social and economic factors should be taken into account in finding candidate locations. The problem can be partially alleviated by employing a large set of candidate locations with different characteristics. However, it leads to large-scale supply chain optimization problems, most of which are computationally intractable (Mohseni, Pishvaee, & Sahebi, 2016).

In recent years, some studies have highlighted the importance of screened candidate locations in reducing total supply chain cost and proposed two-stage models for the strategic design of the biomass supply chain. As illustrated in Fig. 7.3, the first stage selects the most suitable candidate locations based on economic, environmental, and social criteria employing techniques such as geographic information system (GIS), and data envelopment analysis (DEA). The selected locations serve as inputs for the second stage that utilizes mathematical optimization models to design an optimal biomass supply chain network and to determine final locations and other strategic decisions. The techniques that can be applied in the first stage are briefly described in the following subsections, and mathematical models for biomass supply chain design are presented in the next section.

7.3.1 Geographic information system

GIS, which is a powerful tool for collecting, producing, storing, retrieving, analyzing, and displaying spatial data, has been widely used in many sectors of the biofuel supply chain. One of its most important applications is addressing the question of where to locate biomass cultivation sites, storage and preprocessing facilities, biorefineries and other supply chain elements. The idea of using GIS for the identification and evaluation of candidate locations is put forward by Mohseni et al. (2016), who restrict a given study area to a set of suitable locations for microalgae cultivation by a GIS-based site selection method that takes into account a variety of criteria such as slope, land use, solar radiation, ambient temperature, and distance from water and CO_2 sources. The potential locations are then incorporated into the design phase of a microalgae-to-biodiesel supply chain.

To perform site selection analysis by GIS, first, all criteria influencing the location of facilities are defined and their map layers are prepared, as illustrated in Fig. 7.3. Then, various areas on each layer are graded to obtain a suitability value for each pixel on that layer. For example, if annual rainfall is

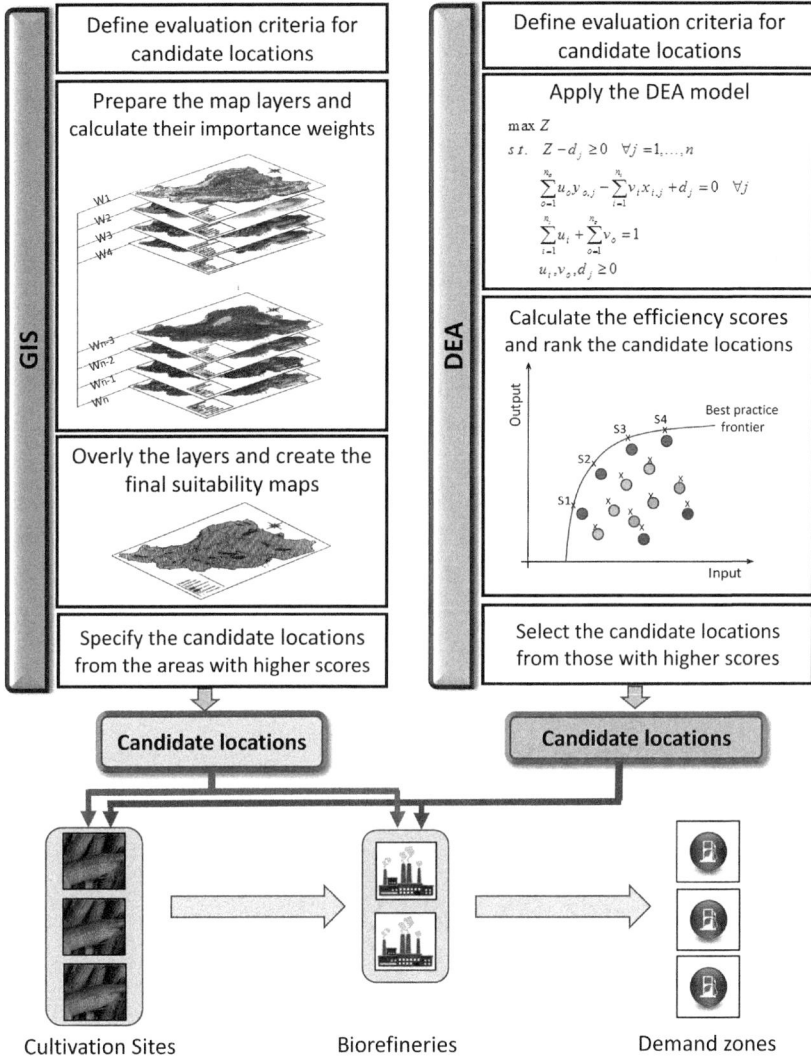

Figure 7.3 *A two-stage framework based on DEA and GIS for candidate locations iden-tification.*

chosen as a criterion to determine the optimal locations for the cultivation of switchgrass, higher values are assigned to pixels associated with areas with higher rainfall and lower values to areas with lower rainfall. In the next step, by one of the multiple criteria decision making (MCDM) methods, the layers are weighted to indicate their relative importance. The layers are then overlaid and aggregated to create the final suitability map. Each pixel of this map is accompanied by a suitability score that is calculated by multiplying the weight assigned to each layer by the value given to the corresponding pixel on that layer and summing the products over all the layers involved (Malczewski, 2004). The higher the score obtained for the pixel, the more appropriate it is for locating the site. Finally, those with higher scores can be designated as the final potential locations.

7.3.2 Data envelopment analysis

DEA, introduced first by Charnes, Cooper, and Rhodes (1978), is a performance measurement methodology for evaluating the relative efficiency of a homogeneous set of decision-making units (DMUs), such as bank branches, hospitals, and firms, which utilize multiple inputs to produce multiple outputs. To benefit from the advantages of this methodology in biomass supply chain design, Babazadeh, Razmi, Rabbani, and Pishvaee (2017) developed a two-stage framework to optimally design a combined Jatropha-waste cooking oil supply chain for biodiesel production. In the first stage, available locations for Jatropha cultivation are assessed by the unified DEA method in which the locations are treated as DMUs. Those that exhibit higher efficiency scores are selected as inputs for the second stage that determines the optimal supply chain design.

With the assumption that there are $n(j=1,...,n)$ DMUS, each representing one of the available locations, the unified DEA model for kth location is formulated as follows (Sueyoshi & Goto, 2011):

$$\max Z_k = \sum_{r=1}^{s} R_r^g D_r^g + \sum_{f=1}^{h} R_f^b D_f^b$$

$$s.t. \sum_{j=1}^{n} g_{r,j} \lambda_j^g - D_r^g = g_{r,k} \quad \forall r = 1,...,s$$

$$\sum_{j=1}^{n} \lambda_j^g = 1$$

$$\sum_{j=1}^{n} b_{f,j} \lambda_j^b + D_r^b = b_{f,k} \quad \forall f = 1,...,h \tag{7.1}$$

$$\sum_{j=1}^{n} \lambda_j^b = 1$$

$$\lambda_j^b, \lambda_j^g, D_r^g, D_f^b \geq 0$$

where the outputs are separated into s desirable (good) outputs $\left(g_{1,j}, g_{2,j}, \ldots, g_{s,j}\right)$ and h undesirable (bad) outputs $\left(b_{1,j}, b_{2,j}, \ldots, b_{h,j}\right)$. $D_r^g (r = 1, \ldots, s)$ and $D_f^b (f = 1, \ldots, h)$ are slack variables related to desirable and undesirable outputs, respectively. $\lambda_j^g (j = 1, \ldots, n)$ and $\lambda_j^b (j = 1, \ldots, n)$ are unknown variables introduced for desirable and undesirable outputs, respectively. These variables, also referred to as intensity or structural variables, connect the outputs by a convex combination. The parameters $R_r^g (r = 1, \ldots, s)$ and $R_f^b (f = 1, \ldots, h)$ that control the ranges of the model are defined as follows:

$$R_r^g = \frac{1}{(s+h+m)\left[\max_j\left\{g_{r,j}\right\} - \min_j\left\{g_{r,j}\right\}\right]} \quad \forall r$$

$$R_j^b = \frac{1}{(s+h+m)\left[\max_j\left\{b_{f,j}\right\} - \min_j\left\{b_{f,j}\right\}\right]} \quad \forall f \qquad (7.2)$$

where m denotes the number of inputs yielding desirable and undesirable outputs. Since it is often difficult to distinguish the inputs from the outputs in site selection analysis by DEA, Babazadeh et al. (2017) import evaluation criteria for Jatropha cultivation site selection into the model as desirable and undesirable outputs, and consider a dummy input (i.e., m is set to 1). They recognize annual rainfall, the annual average of temperature, amount of water resources, available lands, development of agriculture, and population as desirable outputs because locations with higher levels of these criteria have higher priority for Jatropha cultivation, and recognize human development index (HDI) as undesirable outputs because locations situated in cities with lower HDI have higher priority for the development of the biofuel industry to help achieve balanced development throughout the country (Babazadeh, Razmi, Pishvaee, & Rabbani, 2015). After the above model is solved for the kth location, its unified efficiency score is measured as follows:

$$\theta = 1 - \left[\sum_{r=1}^{s} R_r^g D_r^{g\star} + \sum_{f=1}^{h} R_f^b D_f^{b\star}\right] \qquad (7.3)$$

where the superscript ($\star$) denotes the optimality of the slack variables.

Another hybrid model based on DEA and supply chain mathematical modeling was proposed by Mirhashemi, Mohseni, Hasanzadeh, and Pishvaee (2018), who focused on the design and optimization of Moringa oleifera-based biodiesel supply chain considering the capability of Moringa oleifera to combat desertification. All locations threatened by desertification are first ranked by a common weight DEA method and those more at risk are considered as candidate locations to be introduced into the supply chain design model.

While the unified DEA model should be solved for each location (i.e., DMU), the common weight DEA method reduces the number of optimization runs to just one, independently from the number of locations. Moreover, it calculates the efficiency of the DMUs by a set of common importance weights for inputs and outputs, not allowing each DMU to adjust the weights in its favor. The general formulation of the common weight DEA is given as follows (Karsak & Ahiska, 2007):

$$\max Z$$
$$\text{s.t.} \quad Z - d_j \geq 0 \quad \forall j = 1,\ldots,n$$

$$\sum_{o=1}^{n_o} u_o y_{o,j} - \sum_{i=1}^{n_i} v_i x_{i,j} + d_j = 0 \quad \forall j = 1,\ldots,n \tag{7.4}$$

$$\sum_{i=1}^{n_i} u_i + \sum_{o=1}^{n_o} v_o = 1$$

$$u_i, v_o, d_j \geq 0$$

where d_j is the deviation of the jth DMU from the ideal efficiency score of 1, $x_{i,j}$ is the amount of input i utilized by the jth DMU, $y_{o,j}$ is the amount of output o generated by the jth DMU, u_o and v_i are the weights assigned to output o and input i, respectively. n_o and n_i are the numbers of the outputs and inputs, respectively. The objective function minimizes the maximum deviation from efficiency across all the DMUs. Based on the optimal solution of the model, the DMUs with $d_j = 0$ are efficient, and the remaining DMUs that are inefficient can be ranked by calculating their scores as

$$\frac{\sum_o u_o^\star y_{o,j}}{\sum_i v_i^\star x_{i,j}}.$$

Mirhashemi et al. (2018) define the efficiency in the DEA model as being more vulnerable to land degradation and desertification. To identify the locations that are more at risk of these environmental hazards, they consider several site selection factors under two categories of desirable and undesirable criteria. The desirable criteria, which are incorporated into the model as the inputs, increase the risk of desertification when their corresponding values increase in a location, including annual temperature, evaporation, slope gradient, and soil erosion rate. The undesirable criteria, which are incorporated into the model as the outputs, decrease the risk of desertification when their corresponding values decrease, including annual rainfall, the number of actions being taken against desertification, and HDI.

7.4 Biofuel supply chain network design

Supply chain network design, which is also known as strategic supply chain planning, seeks to find the optimal structure of a supply chain by determining strategic level decisions such as the number, location, and capacity of supply chain facilities while simultaneously ensuing the optimal flow of materials throughout the chain (Pishvaee, Rabbani, & Torabi, 2011). Based on a systematic literature review conducted by Ghaderi, Pishvaee, and Moini (2016), the growing shift towards biofuels has led to a surge of interest in developing supply chain design models for biofuel production from various biomass feedstocks, including food crops (Dal-Mas, Giarola, Zamboni, & Bezzo, 2011, Mele, Kostin, Guillén-Gosálbez, & Jiménez, 2011), crop and frost residues (Bairamzadeh, Pishvaee, & Saidi-Mehrabad, 2015; Cambero, Sowlati, Marinescu, & Röser, 2015), dedicated energy crops like switchgrass and jatropha (Zhang, Osmani, Awudu, & Gonela, 2013; Babazadeh et al., 2017), and microalgae (Mohseni & Pishvaee, 2016; Nodooshan et al., 2018).

The biomass supply chain faces unique challenges and is subjected to various sources of uncertainty, which makes designing an efficient supply chain problematic for managers (Bairamzadeh, Saidi-Mehrabad, & Pishvaee, 2018). On the supply side, biomass production is highly variable since biomass yield is affected by a wide range of factors such as climate conditions, farmers planting decisions, plant diseases, and insect infestation (Azadeh & Arani, 2016). The geographical dispersion of biomass sources and the seasonality of biomass are other critical issues governing the supply of biomass. On the production side, most of the technologies for converting biomass into biofuel are still under development and not commercially implemented, which leads to high uncertainty in technical and economic parameters associated with biofuel production (An, Wilhelm, & Searcy, 2011). On the demand side, it is difficult to estimate biofuel demand and price due to the fact that they change as a function of numerous factors such as public acceptance of biofuels, government incentives and regulations, oil price, and other energy market variables.

A well-designed supply chain substantially contributes to reducing the production cost of biofuel, accelerating the transition of the biofuel industry to an economically viable enterprise. Moreover, it is a prerequisite for the effective and efficient management of the biomass-to-biofuel supply chain as the structure and design of the supply chain has a significant influence on tactical and operational-level decisions, thereby impacting on the overall performance of the supply chain (Ekşioğlu, Acharya, Leightley, & Arora, 2009). A biofuel supply chain is generally composed of multiple components in multiple echelons, namely, cultivation sites, storage/preprocessing centers, biorefineries, distribution centers, and demand zones. Biomass resources collected from the cultivations sites are transported to the storage/preprocessing centers where chemical and thermal treatments are unusually applied to biomass for moisture content reduction, contaminants removal, and durability increase (Yue et al., 2014). The biomass is then transported to the biorefineries for converting into biofuel. Finally, the produced biofuel is transported to the demand zones through the distribution centers. Based on the type of biomass and the geographical extent of the supply chain, some of these sites can be combined in integrated facilities or additional sites may be added.

Supply chain network design models developed for biofuel supply chain aim to (1) determine strategic-level decisions with long-lasting effects such as the number, location, and capacity of facilities are to be established and the type of conversion technology of biorefineries, and to (2) provide a strategic plan for the flow of materials, which can include transportation, production, and storage levels (An et al., 2011). To show how biofuel supply chains are designed, a mathematical optimization model for the design of switchgrass-to-bioethanol supply chain under uncertainty is presented in the following, and the applicability of the model is illustrated by a case study in Iran.

7.4.1 Switchgrass-to-bioethanol supply chain design: a case study

The structure of a switchgrass-based bioethanol supply chain is depicted in Fig. 7.4. Switchgrass is cultivated in marginal lands identified as potentially suitable locations for switchgrass cultivations and then harvested by three different harvesting techniques, namely, round bales, square bales, or loose chop (Zhang et al., 2013). If the first two techniques are applied, the collected switchgrass is stored in warehouses adjacent to the cultivation sites or directly transported to the biorefineries. If the last technique is applied, the

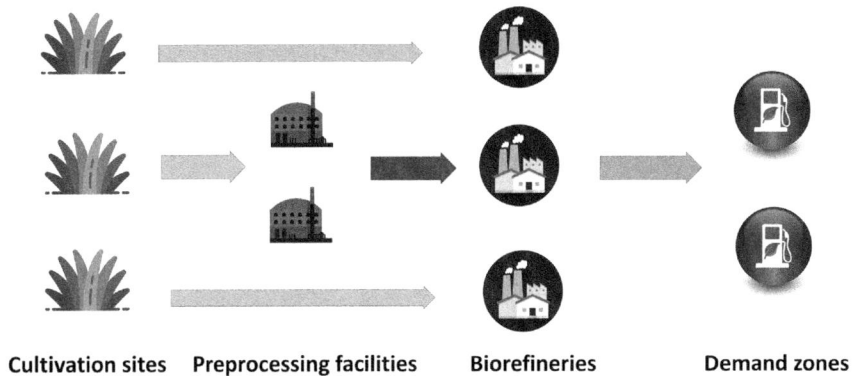

| Cultivation sites | Preprocessing facilities | Biorefineries | Demand zones |

Figure 7.4 *The structure of switchgrass-to-bioethanol supply chain.*

collected switchgrass is transported to the preprocessing facilities for tor-refaction and densification, and then the densified biomass that has higher energy content and durability is transported to the biorefineries when required. The biomass is converted into bioethanol in the biorefineries by either thermochemical or biochemical processing, which are the two main conversion pathways for bioethanol production from lignocellulosic biomass. Finally, the bioethanol is transported to the demand zones.

The variability of the critical factors related to the biofuel supply chain over time, such as biomass yield and quality, could significantly impact the supply chain design (An et al., 2011). Moreover, the demand for biofuel is all-year-round while the supply of biomass is discrete owing to the seasonality (Yue et al., 2014). To capture such dynamics, biofuel supply chain design models are usually formulated as multi-period mathematic models, each period comprises at least 3 months duration, which is justified by the seasonal nature of biomass production. Another important issue is how to determine the length of the planning horizon. Although this is dependent on the decision maker's perspective, the time horizon of 1 year is often decided in the case of biofuel production from food crops and biomass residues that are already available, and it increases to multiple years in the case of some of the dedicated energy crops that take many years to reach maturity after they are planted. As switchgrass requires up to 3 years to achieve its full production potential (Samson, 2007), and it is harvested once a year in the fall after the first killing frost (Zhang et al., 2013), the planning horizon of its supply chain model is assumed to be 3 years with 12 time periods.

Due to changing weather and climate conditions, government policies and regulations, biofuel market structure, and technology, the environment under which the biofuel supply chain is expected to function correctly is highly uncertain. These uncertainties manifest themselves as variations in the parameters of the supply chain design models. In view of the fact that the supply chain design and strategic-level decisions are difficult to change in the short term, they should be immunized against uncertainty to be reliable enough to be used over many years (Pishvaee et al., 2011). As pointed out in the related literature, the switchgrass-based bioethanol supply chain faces various sources of uncertainty (Ghaderi, Moini, & Pishvaee, 2018; Zhang et al., 2013), which can be categorized into (1) uncertainty in switchgrass yield associated with climate variability, (2) uncertainty in the process conversion rate as it is often estimated based on laboratory experiments or pilot plant data, and (3) uncertainty in the fixed and variable costs that arises from the inaccuracies inherent in cost estimates as well as the cost fluctuations that are common in the business environment.

To provide an adequate hedge against these uncertainties, the robust optimization approach is applied in this study because of two main reasons. First, the commercial-scale bioethanol production from switchgrass has not been well established, and as a result, there is no sufficient historical data to determine the probability distribution of uncertain parameters that is required for the stochastic programming approach. In such a situation, robust optimization that only needs the bounds of uncertain parameters is a good choice. Second, robust optimization can make a trade-off between supply chain cost and reliability, making it an appealing tool to provide supply chain decisions that will remain feasible for almost all possible realizations of uncertain parameters.

7.4.1.1 Mathematical model

A multi-period mixed-integer linear program model is developed for the supply chain design problem addressed in the previous section. A list of indices, parameters, and decision variables is given in Table 7.1. Note that the uncertain parameters are distinguished by a tilde. The objective function and constraints of the proposed model are explained in the following.

7.4.1.1.1 Objective function

The objective function minimizes the total cost of the switchgrass-based bioethanol supply chain, which consists of six components: fixed cost, switchgrass production cost, storage cost, preprocessing cost, biofuel production cost, and transportation cost. The objective function and its components are expressed by the following equations:

Table 7.1 Indices, parameters, and decision variables used in the proposed model.

Sets

i	Index of potential locations for switchgrass cultivation $i = (1,\ldots,I)$
p	Index of potential locations for preprocessing facilities $p = (1,\ldots,P)$
b	Index of potential locations for biorefineries $b = (1,\ldots,B)$
c	Index of capacity levels of biorefineries $c = (1,\ldots,C)$
d	Index of demand zones $d = (1,\ldots,D)$
t	Index of time stages $t = (1,\ldots,T)$
h	Index of harvesting method: $h = 1$ (square bales), $h = 2$ (round bales), $h = 3$ (loose chop)
k	Index of conversion technologies: $k = 1$ (thermochemical), $k = 2$ (biochemical)

Parameters

mml_i	Maximum marginal land available in cultivation site i (ha)
$\widetilde{sy}_i^t$	Harvestable yield of switchgrass in cultivation site i at time stage t (ton/ha)
dr_h	Degradation rate of harvested biomass by technique $h \neq 3$ during storage (%)
cap_p	Maximum capacity of preprocessing facility p (ton)
cab_{bc}	Maximum capacity of biorefinery b with capacity level c (L)
$\widetilde{\beta}_{bk}$	conversion rate of densified biomass to ethanol in biorefinery b with technology k (L/ton)
$\widetilde{\alpha}_{hkb}$	conversion rate of harvested biomass by technique $h \neq 3$ to ethanol in biorefinery b with technology k (L/ton)
$\widetilde{bd}_d^t$	Bioethanol demand at demand zone d at time stage t (L)
$\widetilde{fc}_i$	Fixed cost of cultivation site i ($)
$\widetilde{fp}_p$	Fixed cost of preprocessing facility p ($)
$\widetilde{fb}_{bck}$	Fixed cost of biorefinery b with technology k and capacity c ($)
$\widetilde{cc}_i$	Unit cultivation cost of switchgrass in cultivation site i ($/ha)
$\widetilde{scb}_{hi}$	Unit storage cost of harvested biomass by technique $h \neq 3$ in cultivation site I ($/ton)
$\widetilde{scd}_p$	Unit storage cost of densified biomass in preprocessing facility p ($/ton)
$\widetilde{hc}_h$	Unit harvesting cost of switchgrass by technique h ($/ton)
$\widetilde{pc}_p$	Unit processing cost of switchgrass in preprocessing facility p ($/ton)

(Continued)

Table 7.1 Indices, parameters, and decision variables used in the proposed model. *(Cont.)*

$\widetilde{bc}_{bk}$	Unit production cost of bioethanol in biorefinery b with technology k (\$/L)
$\widetilde{th}_{hib}$	Unit transportation cost of harvested biomass by technique $h \neq 3$ from cultivation site i to biorefinery b (\$/ton)
$\widetilde{tl}_{ip}$	Unit transportation cost of loose chopped biomass from cultivation site i to preprocessing facility p (\$/ton)
$\widetilde{td}_{pb}$	Unit transportation cost of densified biomass from preprocessing facility p to biorefinery b (\$/ton)
$\widetilde{tb}_{bd}$	Unit transportation cost of bioethanol from biorefinery b to demand zone d (\$/L)

Positive continuous variables

ml_i	Area of marginal land devoted to switchgrass cultivation in cultivation site i (ha)
aw_{hi}^t	Amount of switchgrass harvested by technique h at time stage t in cultivation site i (ton)
asb_{hi}^t	Amount of harvested biomass by technique $h \neq 3$ stored in cultivation site i at time stage t (ton)
asd_p^t	Amount of densified biomass stored in preprocessing facility p at time stage t (ton)
ah_{hibk}^t	Amount of harvested biomass by technique $h \neq 3$ transported from cultivation site i to biorefinery b with technology k at time stage t (ton)
al_{ip}^t	Amount of loose chopped biomass transported from cultivation site i to preprocessing facility p at time stage t (ton)
ad_{pbk}^t	Amount of densified biomass transported from preprocessing facility p to biorefinery b with technology k at time stage t (ton)
ab_{bkd}^t	Amount of bioethanol produced in biorefinery b with technology k at time stage t and transported to demand zone d (L)

Binary variables

X_i	1 if a cultivation site is built in location i; Else 0
Y_p	1 if a preprocessing facility is built in location p; Else 0
Z_{bck}	1 if a biorefinery with technology k and capacity c is built in location b; Else 0

Total supply chain cost = Fixed cost + Switchgrass productin cost

+ Storage cost + Preprocessing cost + Biofuel production cost

+ Transportation cost   (7.5)

$$\text{Fixed cost}: \sum_i \widetilde{fc}_i X_i + \sum_p \widetilde{fp}_p Y_p + \sum_b \sum_c \sum_k \widetilde{fb}_{bck} Z_{bck} \tag{7.6}$$

$$\text{Switchgrass productin cost}: \sum_i \widetilde{cc}_i ml_i + \sum_h \sum_t \widetilde{hc}_h aw^t_{hi} \tag{7.7}$$

$$\text{Storage cost}: \sum_h \sum_i \sum_t \widetilde{scb}_{hi} asb^t_{hi} + \sum_p \sum_t \widetilde{scd}_p asd^t_p \tag{7.8}$$

$$\text{Preprocessing cost}: \sum_i \sum_p \sum_t \widetilde{pc}_p al^t_{ip} \tag{7.9}$$

$$\text{Biofuel production cost}: \sum_b \sum_c \sum_k \sum_t \sum_d \widetilde{bc}_{bk} ab^t_{bkd} \tag{7.10}$$

$$\begin{aligned}\text{Transportation cost}: &\sum_h \sum_i \sum_b \sum_k \sum_t \widetilde{th}_{hib} ah^t_{hibk} + \sum_i \sum_p \sum_t \widetilde{tl}_{ip} al^t_{ip} \\ &+ \sum_p \sum_b \sum_k \sum_t \widetilde{td}_{pb} ad^t_{pbk} + \sum_b \sum_d \sum_k \sum_t \widetilde{tb}_{bd} ab^t_{bkd}\end{aligned} \tag{7.11}$$

7.4.1.1.2 Constraints
- Cultivation sites

Constraint (7.12) ensures that, in each cultivation site (if built in location i), the area of marginal land devoted to switchgrass cultivation does not exceed the maximum available marginal land.

$$ml_i \leq mml_i X_i \quad \forall i \tag{7.12}$$

Constraint (7.13) ensures that, in each cultivation site and at time stage, the sum of the harvested switchgrass as round bales, square bales, and loose chop is not greater than the amount of switchgrass that can be harvested.

$$\sum_h aw^t_{hi} \leq \widetilde{sy}^t_i ml_i \quad \forall i, t \tag{7.13}$$

Constraint (7.14) ensures that, in each cultivation site and at each time stage, all the loose chopped switchgrass is sent to the preprocessing facilities.

$$\sum_p al^t_{ip} = aw^t_{hi} \quad \forall i, t, h = 3 \tag{7.14}$$

Constraint (7.15) ensures that, for each cultivation site, the amount of the harvested switchgrass as round or square bales plus the inventory at left in the previous stage (considering the deterioration rate) is equal to the round or square bales sent to the biorefineries plus the inventory at the end of the current stage.

$$asb_{hi}^{t-1}\left(1 - dr_h\right) + aw_{hi}^t = asb_{hi}^t + \sum_b \sum_k ah_{hibk}^t \quad \forall i, t, \quad {\scriptstyle h=1,2}$$ (7.15)

- Preprocessing facilities

Constraint (7.16) ensures that, at each time stage, the amount of loose chopped biomass sent to each preprocessing facility (if built in location p) is not greater than its capacity

$$\sum_i al_{ip}^t \leq cap_p Y_p \quad \forall p, t$$ (7.16)

Constraint (7.17) ensures that, in each preprocessing facility, the amount of the switchgrass coming from the cultivation sites plus the inventory at left in the previous stage is equal to the densified biomass sent to the biorefineries plus the inventory at the end of the current stage.

$$asd_p^{t-1} + \sum_i al_{ip}^t = asd_p^t + \sum_b \sum_k ad_{pbk}^t \quad \forall p, t$$ (7.17)

- Biorefineries

Constraint (7.18) ensures that, in each biorefinery and at each time stage, the total amount of bioethanol sent to the demand zones is not greater than the sum of the flows of the switchgrass coming from the cultivation sites and preprocessing facilities multiplied by the corresponding conversion rate.

$$\sum_p \widetilde{\beta_{kb}} ad_{pbk}^t + \sum_h \sum_i \widetilde{\alpha_{hkb}} ah_{hibk}^t \geq \sum_d ab_{bkd}^t \quad \forall b, k, t$$ (7.18)

Constraint (7.19) ensures that, at each time stage, the amount of bioethanol production in each biorefinery does not exceed its chosen capacity.

$$\sum_d ab_{bkd}^t \leq cab_{bc} Z_{bck} \quad \forall b, c, k, t$$ (7.19)

Constraint (7.20) ensures that only one capacity level and one technology type are selected for each biorefinery.

$$\sum_c \sum_k Z_{bck} \leq 1 \quad \forall b$$ (7.20)

- Demand zones

Constraint (7.21) ensures that, in each demand zone and at each time stage, the bioethanol demand is satisfied by the amount of bioethanol coming from the biorefineries.

$$\sum_b \sum_k ab^t_{bkd} \geq \widetilde{bd}^t_d \quad \forall d,t \tag{7.21}$$

7.4.1.2 Robust counterpart formulation

In this section, the above deterministic supply chain design model is extended to a robust counterpart model, taking into account uncertainty in switchgrass yield, cultivation and harvesting cost, switchgrass preprocessing and bioethanol production cost, process yield (i.e., switchgrass to bioethanol conversion rate), storage cost, transportation cost, and bioethanol demand. To do so, the combined box and polyhedral set-induced robust counterpart, which is exactly similar to the budgeted robust counterpart developed by Bertsimas and Sim (2004), is adopted. This robust formulation provides great flexibility in controlling the conservatism degree of the robust solution while preserving the linear structure of the original problem. In the robust optimization framework, each uncertain parameter is modeled as a random variable that is assumed to be symmetrically distributed in a perturbation range around its nominal value. For example, $\widetilde{cr}_k$ belongs to the range $\left[cr_k - \widehat{cr}_k, cr_k + \widehat{cr}_k \right]$, where cr_k is the nominal value and $\widehat{cr}_k$ is the perturbation amplitude. With this assumption, the robust counterparts of the constraints and objective function are derived in the following.

7.4.1.2.1 Constraints

Following the procedure explained in Chapter 6, constraint (7.18) is first transformed into the following nonlinear robust counterpart:

$$\sum_p \beta_{kb} ad^t_{pbk} + \sum_h \sum_i \alpha_{hkb} ah^t_{hibk}$$

$$- \max \left(\sum_p \widehat{\beta}_{kb} ad^t_{pbk} Z_{kbt} + \sum_h \sum_i \widehat{\alpha}_{hkb} ah^t_{hibk} Z_{hkbt} \right) \geq \sum_d ab^t_{bkd} \quad \forall b,k,t$$

$$\text{s.t. } \sum_h Z_{hkbt} + Z_{kbt} \leq \Gamma_{kbt} \quad \forall k,b,t$$

$$0 \leq Z_{hkbt} \leq 1 \quad \forall h,k,b,t$$

$$0 \leq Z_{kbt} \leq 1 \quad \forall k,b,t$$

(7.22)

For each constraint row, the uncertainty budget Γ_{kbt} is introduced, which takes values in the interval $[0,3]$. The robust formulation forces Γ_{kbt} of the uncertain coefficients to take their worst-case values and an additional coefficient to change slightly. Replacing the inner maximization with its corresponding dual form, the flowing robust counterpart is obtained:

$$\sum_p \beta_{kb} ad^t_{pbk} + \sum_h \sum_i \alpha_{hkb} ah^t_{hibk}$$

$$-\omega^t_{bk} - \sum_h \mu^t_{hbk} - \Gamma_{bkt}\lambda_{bkt} \geq \sum_d ab^t_{bkd} \quad \forall b,k,t$$

$$\lambda_{bkt} + \omega^t_{bk} \geq \sum_p \hat{\beta}_{kb} ad^t_{pbk} \quad \forall b,k,t \tag{7.23}$$

$$\lambda_{bkt} + \mu^t_{hbk} \geq \sum_i \hat{\alpha}_{hkb} ah^t_{hibk} \quad \forall b,k,t,h$$

$$\omega^t_{bk}, \mu^t_{hbk}, \lambda_{bkt} \geq 0$$

where $\omega^t_{bk}, \mu^t_{hbk}$ and λ_{bkt} are dual variables.

For constraints (7.13) and (7.21) including one uncertain coefficient, the uncertainty budges Γ'_{it} and Γ''_{dt} with the range of $[0,1]$ are introduced, and the original constrains are reformulated as follows:

$$\sum_h aw^t_{hi} \leq sy^t_i ml_i - \Gamma'_{it} \widehat{sy^t_i} ml_i \quad \forall i,t \tag{7.24}$$

$$\sum_b \sum_k ab^t_{bkd} \geq bd^t_d + \Gamma''_{dt} \widehat{bd^t_d} \quad \forall d,t \tag{7.25}$$

7.4.1.2.2 Objective function

To keep the notation manageable in the sequel, the objective function is first represented in the following compact form:

$$\min \quad \sum_m \widetilde{c}_m x_m + \sum_n \widetilde{f}_n y_n \tag{7.26}$$

where x and y denote the binary and continuous variables, respectively, and c and f correspond to the fixed and variable cost parameters, respectively. To treat the objective function as the uncertain constraints, it is equivalently transformed as follows:

$$\min \quad r$$
$$s.t. \sum_m \widetilde{c}_m x_m + \sum_n \widetilde{f}_n y_n \leq r \tag{7.27}$$

Now, the robust counterpart of the above problem can be formulated as follows:

$$\min \quad r$$

$$s.t. \sum_m c_m x_m + \sum_n f_n y_n + \sum_m p_m + \sum_n q_n + \Gamma''' z \leq r$$

$$z + p_m \geq \widehat{c}_m x_m \quad \forall m$$

$$z + q_n \geq \widehat{f}_n y_n \quad \forall n \tag{7.28}$$

where p_m, q_n and z are dual variables, and Γ''' is the uncertainty budget that takes values in the range $[0, N]$, where N is the number of uncertain parameters in the objective function.

7.4.1.3 Case study

To demonstrate the application of the proposed model, it is utilized to design a switchgrass-based bioethanol supply chain in Iran. There are several motivations for bioethanol production from switchgrass in Iran. First, developing renewable energy and diversifying the energy mix help lessen the dependence on fossil fuel and the associated environmental problems. Second, the cultivation of plants like switchgrass that can grow with low water consumption avoids the concerns about the negative impact of biofuel production on water resources in Iran. Third, Iran has large areas of arid and semi-arid lands, which if exploited for switchgrass production, can improve soil organic matter, control erosion, and promote economic growth in rural areas (Jensen et al., 2007).

Given the fact that switchgrass needs to be grown under specific conditions, identifying the candidate locations for switchgrass cultivation is of great importance before implementing the supply chain design model. As switchgrass cannot tolerate extreme chilling and low soil PH (below 5.0), spatial filtering is performed by GIS to omit the areas not satisfying these critical from the available marginal lands in Iran. From the remaining areas, 30 places are selected as initial candidate locations. Using a cross–efficiency DEA model developed by (Khanjarpanah, Pishvaee, & Seyedhosseini, 2017), these locations are ranked according to different criteria such as land cost, annual precipitation, average temperature, water resources availability, and farming development. Finally, 15 locations with the highest suitability score are determined as final candidate locations. Iran is comprised of 30 provinces, each of which is considered as a node in the model, representing a possible preprocessing facility, a potential biorefinery, and a demand zone. It is assumed that there are one capacity level for the preprocessing facilities (900,000 ton) and four capacity levels for the biorefineries (50, 100, 150, and 200 million liters (ML)). The values of other input parameters can be found in Ghaderi et al. (2018).

7.4.1.3.1 Results and discussion

The optimal supply chain design for the proposed deterministic model is shown in Fig. 7.5. Among the 15 candidate locations for switchgrass cultivation, 9 locations are determined as the optimal locations for the cultivation sites, whose devoted marginal land areas range from 11124.2 to 28745.6 ha. Having large areas of marginal lands, low land cost, and high switchgrass yield appear to be the main reason why these locations are chosen. The model prescribes five preprocessing facilities, all of which are located in the vicinity of the cultivation sites to avoid transporting low–density switchgrass for long distances. In all the cultivation sites, harvesting switchgrass as loose chop is preferred to square or round bales, despite the fact that it adds the costs of the preprocessing facilities to the supply chain cost. The model also

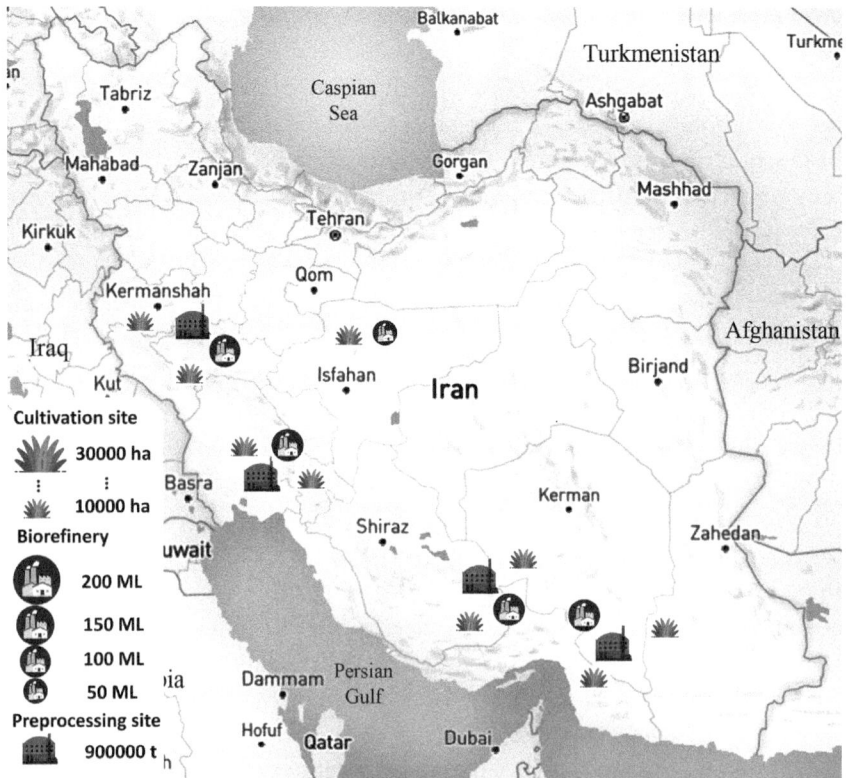

Figure 7.5 *The optimal supply chain design determined by the deterministic model.*

selects five biochemical biorefineries, one of which has the capacity of 50 ML and the others have the capacity of 100 ML.

To compare the supply chain design obtained from the robust model with that from the deterministic model, two different settings for the robust model are considered, namely, hard worst-case and realistic worst-case. The hard worst-case version assumes a variation range of 20% for the uncertain parameters to ensure that the robust solution remains feasible even if the uncertain parameters vary by 20% from their nominal values, and also sets each uncertainty budget to the maximum of its corresponding range to avoid the possibility of constraints violation for all realizations of the uncertain parameters. Although this version provides maximum protection against uncertainty, it leads to a high cost of robustness that manifests itself in the form of a significant deterioration in the objective function value. To make a trade-off between the robustness and its cost, the realistic worst-case version considers a 10% variation in the uncertain parameters and also sets each uncertainty budget to the middle of its corresponding range, which allows the constraints violation for some realizations of the uncertain parameters.

Comparing the results in Figs. 7.5-7.7, the determinist and robust models determine almost the same locations for the cultivation sites, preprocessing facilities and biorefineries as well as the same conversion technology for the biorefineries. This indicates the robustness of these strategic decisions in the face of uncertainty, meaning that they are not invalidated by small variations in the input parameters. However, this is not true for the capacity decisions. Compared to the determinist model, the robust model suggests fewer but larger cultivations sites and biorefineries, providing higher total capacity. This is more evident in the hard worst-case version that is capable of satisfying the constraints even when the switchgrass yield, bioethanol demand, and other uncertain parameters reach their worst-case value, simultaneously.

Generally, robust supply chain design models enhance the reliably of capacity decisions by determining more facilities (i.e., a more distributed supply chain structure) or fewer with higher capacities (i.e., a more centralized supply chain structure). The centralized structure incurs higher transportation costs and lower capital costs than the distributed structure (Mirhashemi et al., 2018). This is because the centralized structure can benefit more from economies of scale but it requires transporting bulky biomass for longer distances from more scattered supply sites to centralized biorefineries. In this case study, as the conservatism level of the robust model increases, the structure of the switchgrass-based bioethanol supply chain becomes more

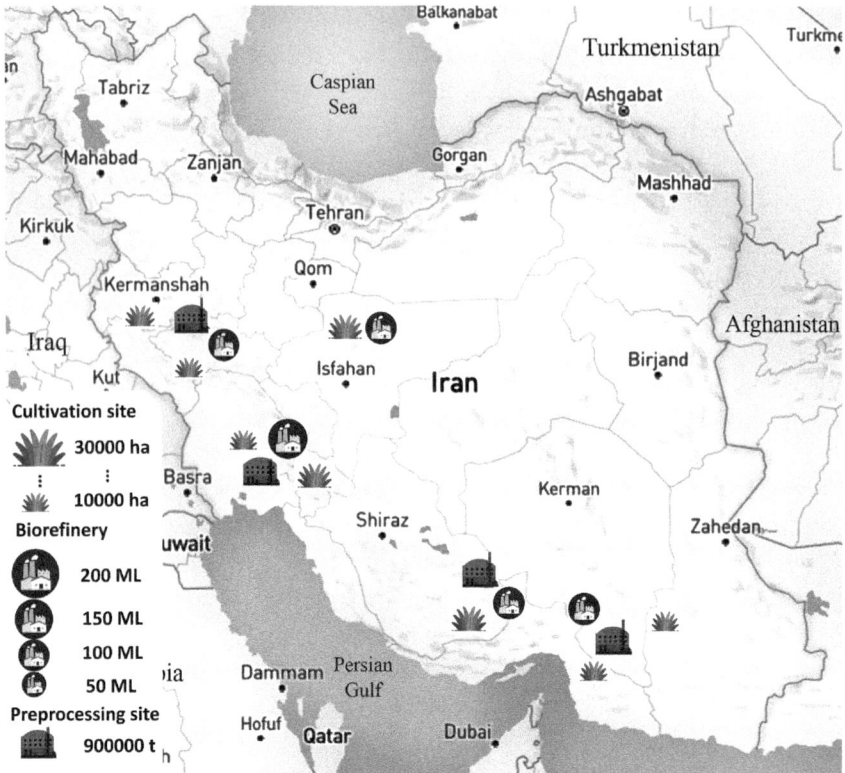

Figure 7.6 *The optimal supply chain design determined by the realistic worst-case robust version.*

centralized. The reason is that the effect of variations in the fixed cost parameters on the supply chain cost outweighs that of variations in the transportation cost parameters, and the robust model determines fewer but larger facilities that are more cost-effective because of economies of scale, thereby attenuating the adverse effect of uncertainty in the fixed costs.

7.5 Conclusions

The strategic network design of the biomass supply chain represents the long-term decisions made in the different stages of the chain, including, but not limited to, selection of biomass types, biomass and biofuel transportation network design, and determining the number, location, technology and capacity of biorefineries. To show how to model and design biofuel supply chains, a two-stage framework is proposed in this chapter. In the first stage,

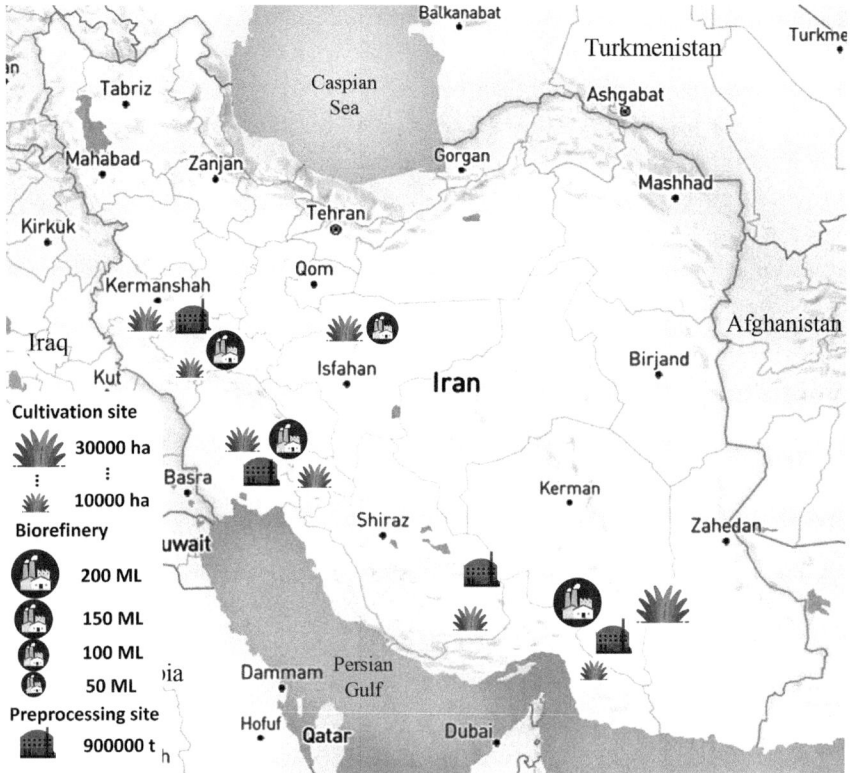

Figure 7.7 *The optimal supply chain design determined by the hard worst-case robust version.*

GIS or DEA is utilized to find the most suitable candidate locations for biorefineries and other supply chain facilities, based upon which the structure of biofuel supply chain is to be designed. The second stage develops a mathematical model to systematically design and optimize the entire switchgrass-to-bioethanol supply chain. The deterministic supply chain design model is then extended to a robust counterpart model, taking into account uncertainty in switchgrass yield, cultivation and harvesting cost, switchgrass preprocessing and bioethanol production cost, process yield (i.e., switchgrass to bioethanol conversion rate, storage cost, transportation cost, and bioethanol demand). To deal with uncertain parameters, the combined box and polyhedral set–induced robust counterpart model is adopted. Comparing the supply chain design obtained from the robust model with that from the deterministic model demonstrates that the deterministic and robust models determine almost the same locations for the cultivation sites, preprocessing facilities, and biorefineries as well as the same conversion technology for the biorefineries. This indicates the robustness of these strategic decisions in the face of uncertainty, meaning that they are not invalidated by small variations in the input parameters. However, compared to the determinist model, the robust model suggests fewer but larger cultivations sites and biorefineries, providing higher total capacity. This is more evident in the hard worst-case version that is capable of satisfying the constraints even when the switch-grass yield, bioethanol demand, and other uncertain parameters reach their worst-case value, simultaneously.

References

An, H., Wilhelm, W. E., & Searcy, S. W. (2011). A mathematical model to design a lignocel-lulosic biofuel supply chain system with a case study based on a region in Central Texas. *Bioresource Technology, 102*, 7860–7870.

Azadeh, A., & Arani, H. V. (2016). Biodiesel supply chain optimization via a hybrid system dynamics-mathematical programming approach. *Renewable Energy, 93*, 383–403.

Babazadeh, R., Razmi, J., Pishvaee, M. S., & Rabbani, M. (2015). A non-radial DEA model for location optimization of *Jatropha curcas* L. cultivation. *Industrial Crops and Products, 69*, 197–203.

Babazadeh, R., Razmi, J., Rabbani, M., & Pishvaee, M. S. (2017). An integrated data envelop-ment analysis-mathematical programming approach to strategic biodiesel supply chain network design problem. *Journal of Cleaner Production, 147*, 694–707.

Bairamzadeh, S., Pishvaee, M. S., & Saidi-Mehrabad, M. (2015). Multiobjective robust pos-sibilistic programming approach to sustainable bioethanol supply chain design under multiple uncertainties. *Industrial & Engineering Chemistry Research, 55*, 237–256.

Bairamzadeh, S., Saidi-Mehrabad, M., & Pishvaee, M. S. (2018). Modelling different types of uncertainty in biofuel supply network design and planning: A robust optimization ap-proach. *Renewable Energy, 116*, 500–517.

Bertsimas, D., & Sim, M. (2004). The price of robustness. *Operations Research, 52*, 35–53.

Cambero, C., Sowlati, T., Marinescu, M., & Röser, D. (2015). Strategic optimization of forest residues to bioenergy and biofuel supply chain. *International Journal of Energy Research, 39*, 439–452.

Charnes, A., Cooper, W. W., & Rhodes, E. (1978). Measuring the efficiency of decision mak-ing units. *European Journal of Operational Research, 2*, 429–444.

Cobuloglu, H. I., & Büyüktahtakin, I. E. (2014). *Multi-criteria approach for biomass crop selection under fuzzy environment*. IIE Annual Conference Proceedings, Institute of Industrial and Systems Engineers (IISE).

Dal-Mas, M., Giarola, S., Zamboni, A., & Bezzo, F. (2011). Strategic design and investment capacity planning of the ethanol supply chain under price uncertainty. *Biomass and Bio-energy, 35*, 2059–2071.

Ekşioğlu, S. D., Acharya, A., Leightley, L. E., & Arora, S. (2009). Analyzing the design and management of biomass-to-biorefinery supply chain. *Computers & Industrial Engineering, 57*, 1342–1352.

Ghaderi, H., Moini, A., & Pishvaee, M. S. (2018). A multi-objective robust possibilistic pro-gramming approach to sustainable switchgrass-based bioethanol supply chain network design. *Journal of Cleaner Production, 179*, 368–406.

Ghaderi, H., Pishvaee, M. S., & Moini, A. (2016). Biomass supply chain network design: An optimization-oriented review and analysis. *Industrial Crops and Products, 94*, 972–1000.

Hu, H., Lin, T., Wang, S., & Rodriguez, L. F. (2017). A cyberGIS approach to uncertainty and sensitivity analysis in biomass supply chain optimization. *Applied Energy, 203*, 26–40.

Jensen, K., Clark, C. D., Ellis, P., English, B., Menard, J., Walsh, M., & De La Torre Ugarte, D. (2007). Farmer willingness to grow switchgrass for energy production. *Biomass and Bioenergy, 31*, 773–781.

Karsak, E. E., & Ahiska, S. S. (2007). A common-weight MCDM framework for decision problems with multiple inputs and outputs. In *International Conference on Computational Science and Its Applications* (pp. 779–790). Springer.

Kashanian, M., Pishvaee, M. S., & Sahebi, H. (2020). Sustainable biomass portfolio sourcing plan using multi-stage stochastic programming. *Energy*, 117923.

Khanjarpanah, H., Pishvaee, M. S., & Seyedhosseini, S. M. (2017). A risk averse cross-efficiency data envelopment analysis model for sustainable switchgrass cultivation location optimization. *Industrial Crops and Products, 109*, 514–522.

Li, Q., & Hu, G. (2014). Supply chain design under uncertainty for advanced biofuel production based on bio-oil gasification. *Energy, 74*, 576–584.

Malczewski, J. (2004). GIS-based land-use suitability analysis: A critical overview. *Progress in Planning, 62*, 3–65.

Marufuzzaman, M., & Ekşioğlu, S. D. (2016). Designing a reliable and dynamic multimodal transportation network for biofuel supply chains. *Transportation Science, 51*, 494–517.

Marufuzzaman, M., Eksioglu, S. D., & Huang, Y. (2014). Two-stage stochastic programming supply chain model for biodiesel production via wastewater treatment. *Computers & Operations Research, 49*, 1–17.

Marvin, W. A., Schmidt, L. D., Benjaafar, S., Tiffany, D. G., & Daoutidis, P. (2012). Economic optimization of a lignocellulosic biomass-to-ethanol supply chain. *Chemical Engineering Science, 67*, 68–79.

Mele, F. D., Kostin, A. M., Guillén-Gosálbez, G., & Jiménez, L. (2011). Multiobjective model for more sustainable fuel supply chains. A case study of the sugar cane industry in Argentina. *Industrial & Engineering Chemistry Research, 50*, 4939–4958.

Melo, M. T., Nickel, S., & Saldanha-Da-Gama, F. (2009). Facility location and supply chain management-A review. *European Journal of Operational Research, 196*, 401–412.

Mirhashemi, M. S., Mohseni, S., Hasanzadeh, M., & Pishvaee, M. S. (2018). Moringa oleifera biomass-to-biodiesel supply chain design: An opportunity to combat desertification in Iran. *Journal of Cleaner Production, 203*, 313–327.

Mohseni, S., & Pishvaee, M. S. (2016). A robust programming approach towards design and optimization of microalgae-based biofuel supply chain. *Computers & Industrial Engineering, 100*, 58–71.

Mohseni, S., Pishvaee, M. S., & Sahebi, H. (2016). Robust design and planning of microalgae biomass-to-biodiesel supply chain: A case study in Iran. *Energy, 111*, 736–755.

Nodooshan, K. G., Moraga, R. J., Chen, S.-J. G., Nguyen, C., Wang, Z., & Mohseni, S. (2018). Environmental and economic optimization of algal biofuel supply chain with multiple technological pathways. *Industrial & Engineering Chemistry Research, 57*, 6910–6925.

Pérez, A. T. E., Camargo, M., Rincón, P. C. N., & Marchant, M. A. (2017). Key challenges and requirements for sustainable and industrialized biorefinery supply chain design and management: A bibliographic analysis. *Renewable and Sustainable Energy Reviews, 69*, 350–359.

Pishvaee, M. S., Rabbani, M., & Torabi, S. A. (2011). A robust optimization approach to closed-loop supply chain network design under uncertainty. *Applied Mathematical Modelling, 35*, 637–649.

Samson, R. (2007). *Switchgrass production in Ontario: A management guide*. Resource Efficient Agricultural Production (REAP)-Canada.

Sueyoshi, T., & Goto, M. (2011). Measurement of returns to scale and damages to scale for DEA-based operational and environmental assessment: How to manage desirable (good) and undesirable (bad) outputs? *European Journal of Operational Research*, *211*, 76–89.

Sun, A., Davis, R., Starbuck, M., Ben-Amotz, A., Pate, R., & Pienkos, P. T. (2011). Comparative cost analysis of algal oil production for biofuels. *Energy*, *36*, 5169–5179.

Tong, K., Gong, J., Yue, D., & You, F. (2013). Stochastic programming approach to optimal design and operations of integrated hydrocarbon biofuel and petroleum supply chains. *ACS Sustainable Chemistry & Engineering*, *2*, 49–61.

Yatim, P., Lin, N. S., Lam, H. L., & Choy, E. A. (2017). Overview of the key risks in the pioneering stage of the Malaysian biomass industry. *Clean Technologies and Environmental Policy*, *19*, 1825–1839.

You, F., & Wang, B. (2011). Life cycle optimization of biomass-to-liquid supply chains with distributed-centralized processing networks. *Industrial & Engineering Chemistry Research*, *50*, 10102–10127.

Yue, D., & You, F. (2016). Optimal supply chain design and operations under multi-scale uncertainties: Nested stochastic robust optimization modeling framework and solution algorithm. *AIChE Journal*, *62*, 3041–3055.

Yue, D., You, F., & Snyder, S. W. (2014). Biomass-to-bioenergy and biofuel supply chain optimization: Overview, key issues and challenges. *Computers & Chemical Engineering*, *66*, 36–56.

Zhang, J., Osmani, A., Awudu, I., & Gonela, V. (2013). An integrated optimization model for switchgrass-based bioethanol supply chain. *Applied Energy*, *102*, 1205–1217.

CHAPTER 8

Tactical planning in biofuel supply chain under uncertainty

8.1 Introduction

Tactical decisions in the supply chain are medium-term decisions made over a planning period ranging from a few months to one year. These decisions provide an outline of the regular processes and operations of the supply chain, namely procurement, production, distribution, and sales according to the supply structure established in the strategic-level decision making at the beginning of the planning horizon (i.e., at time zero). Tactical decision planning aims to bridge the gap between the strategic long-term decisions and operational short-term decisions relating to the day-to-day operations of the supply chain (Church, Murray, & Barber, 2000).

An overview of tactical decisions associated with biomass-to-biofuel supply chains is shown in Fig. 8.1. In the upstream of the supply chain, biomass harvest and collection decisions play a vital role in ensuring that the required quantity of biomass is supplied in a cost-effective and timely manner. Biomass collection is related to the procurement of the required biomass from farms. It may extend to include harvest operations such as mowing, raking, and baling when biomass is not readily available and it must be harvested prior to the collection. Since the harvest and collection of biomass should be completed within tight time windows, it is regarded as one of the crucial tactical decisions that, if not planned properly, can threaten the reliability of biomass supply. These decisions are taken at both medium- and short-term planning levels. For the medium-term level, the decisions include biomass suppliers selections and determining the quantity of biomass is to be harvested and collected from each supplier while the short-term level schedules all harvest and collection operations (Malladi & Sowlati, 2018). Tactical decisions related to biomass storage include the quantity of biomass to be stored at each month or season, as well as decisions a longer lasting effect such as the type of storage system. Due to the specific characteristics of biomass sources such as the seasonal availability, scattered geographical distribution, and uncertain nature of biomass supply arising mainly from unpredictable weather conditions, it is essential to

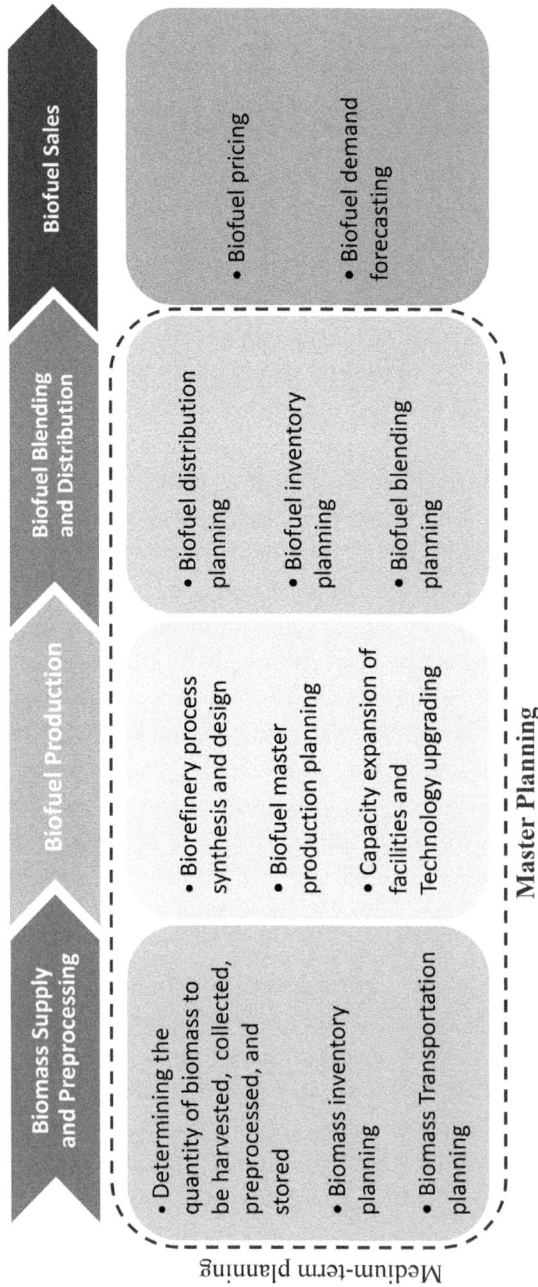

Biofuel Sales

- Biofuel pricing
- Biofuel demand forecasting

Biofuel Blending and Distribution

- Biofuel distribution planning
- Biofuel inventory planning
- Biofuel blending planning

Biofuel Production

- Biorefinery process synthesis and design
- Biofuel master production planning
- Capacity expansion of facilities and Technology upgrading

Biomass Supply and Preprocessing

- Determining the quantity of biomass to be harvested, collected, preprocessed, and stored
- Biomass inventory planning
- Biomass Transportation planning

Master Planning

Medium-term planning

Figure 8.1 *An overview of tactical decisions in biofuel supply chains.*

store sufficient amounts of biomass at different periods of the year to ensure an all-year-round supply of biomass to biorefineries. In addition to the direct storage of biomass, various pretreatment processes such as chemical and thermal treatments can be applied depending on the type of biomass and the corresponding harvesting method, which densifies raw biomass, removes contaminants, reduces the degradation, and prepares it for the conversion processes at biorefineries (Zandi Atashbar, Labadie, & Prins, 2018). Specifying the quantity of biomass to be preprocessed in each period along with the selection of the pretreatment method, are preprocessing decisions that can be incorporated in the medium-term planning of biomass supply chains. As another part of tactical decision-making in the upstream, biomass transportation planning determines the quantity of biomass feedstock to be transported from supply areas to biorefineries in each period as well as the routing of trucks for satisfying the determined transportation orders. The importance of the transportation decisions stems from the fact that for commercial-scale biofuel production, large quantities of low-density biomass resources must be hauled over long distances, making transportation cost a major contributor to the cost of biomass supply chains (Yue, You, & Snyder, 2014).

The midstream of the biomass supply chain contains various tactical decisions ranging from inventory management and production planning to decisions with a longer lasting effect such as the design, synthesis, and upgrading of conversion processes. Storing large quantities of biomass at farms is impossible because of limited storage space. Moreover, biomass storage at farms is time-constrained since they must be prepared for the next planting season (Ebadian, Sokhansanj, & Webb, 2017). Due to these limitations, biomass can be stored at biorefineries or intermediate storage centers located between biomass farms and biorefineries. The dry matter losses and quality deterioration of biomass during storage necessitate developing supply chain planning models with monthly or seasonal time intervals that address the inventory control decisions. The biomass deterioration rates caused by different storage technologies must be evaluated by inventory models to quantify the tradeoff between the cost of storage and the cost of material loss (An, Wilhelm, & Searcy, 2011). Production planning decisions determine the optimal quantity of material produced or consumed in various conversion steps while ensuring the efficient allocation of resources, human labor, and equipment units to fulfill production functions in a timely and economical manner. Long-term tactical decisions that are associated with biorefineries include the design, synthesis, and upgrading of conversion processes, which can be handled by superstructure-based optimization models that determine optimal processing pathways and production-related decisions considering available biomass feedstocks, potential technologies alternatives, and desired final products. Further down the supply chain, on the demand side, biofuel pricing and demand forecasting are among other types of significant medium-term decisions of biofuel supply chains.

Biomass-to-biofuel supply chains are exposed to a wide range of uncertainties and risks arising from issues such as technology evolution, changing policies and regulations, demand and price variability, unpredictable weather conditions, production cost variations, as well as man-made and natural disasters (Bairamzadeh, Saidi-Mehrabad, & Pishvaee, 2018). In the upstream of the biofuel supply chain, biomass yield that is typically subjected to seasonal variations is associated with a high level of uncertainty stemming from several factors such as variability of weather and climate factors (e.g., sunlight, pest threats, and precipitation), and natural catastrophic events (e.g., floods, earthquakes, hurricanes, landslides, and droughts). In the midstream, estimating the biomass conversion rates in producing biofuels is tainted with uncertainty since most of the technologies for converting biomass into biofuel are still under development and not commercially implemented. Biofuel demand is one of the important sources of the downstream uncertainties while the time range over which it fluctuates is driven by numerous factors, such as changing regulations and policies, different regional economic structures and consumption behavior, increasing competitive pressure, and market complexities. Another source of the downstream uncertainties corresponds to biofuel price, which is mainly correlated to fossil fuel prices since the price of crude oil tends to lead the biofuel price.

As discussed in Chapter 7, biofuel supply chain uncertainties can be categorized into strategic (long-term) and operational (short-term) uncertainties based on their degree and type. While operational uncertainties such as biomass yield, biofuel demand and price, and biomass-to-biofuel conversion rates change over a short period of time, strategic uncertainties have long-lasting impacts and remain unchanged for a long time (Yue & You, 2016). Due to the fact that tactical decisions cover the planning horizon from a few months to a year, they are threatened by both strategic and operational uncertainties. Failure to hedge against these uncertainties may result in suboptimal or even infeasible supply chain decisions. Based on the available data associated with uncertain parameters, an appropriate approach must be employed to capture such dynamics. Scenario-based stochastic programming can be employed for situations that uncertain parameters are described by a set of discrete scenarios, while the probability of occurrence associated with each scenario is known. However, when it is impossible to extract probability distributions due to limited or unreliable data, subjective probability distributions can be estimated based on experts' judgments. In these situations, fuzzy programming can treat epistemic uncertainty arising from the lack of complete knowledge about the parameters. On the other hand, when available data is so limited such that objective or subjective probability distribution cannot be obtained, and only the bounds of uncertain parameters can be estimated, set-based robust optimization can be applied. This approach only requires the lower and upper bounds of the uncertain parameters instead of detailed probabilistic knowledge.

8.2 Biorefinery process synthesis and design

As discussed in Chapter 7, supply chain network design models seek to systematically design and optimize the entire biomass-to-biofuel supply chain from biomass supply sites to biofuel end-users, enabling the decision maker to make optimal strategic decisions regarding the location, capacity and technology type of supply chain facilities. An integrated biorefinery can convert multiple raw materials to multiple products through several production routes, each of which consists of a series of processing steps with various technology alternatives. To handle such a complex system, there is a need to develop a systematic approach that investigates the biorefinery and its processing units in a more detailed way than the supply chain design models, which is the subject of this section.

Superstructure-based optimization is a powerful approach to identify and optimize processing units and their connections in an integrated biorefinery, thereby providing the optimal biorefinery configuration (Bao, Ng, Tay, Jiménez-Gutiérrez, & El-Halwagi, 2011). This approach postulates a superstructure that embraces all feasible options for the processing units and all possible interconnections between the units with the purpose of eliminating unfeasible process flowsheets. The postulated superstructure is then translated into a mathematical programming problem, and uncertainties inherent in biorefinery processes are incorporated into the problem with the help of optimization techniques for dealing with uncertainty such as robust optimization and stochastic programming. By solving the superstructure-based optimization model, all the process configurations are assessed, and the optimal process flowsheet along with its operating conditions and design parameters are determined (Mencarelli, Chen, Pagot, & Grossmann, 2020).

The problem of process synthesis and design of biorefinery includes both strategic and tactical decisions, but its strategic decisions are more likely to need revision than those described in Chapter 7 (such as location and capacity of facilities). The reason is that biomass conversion technologies are evolving and the superstructure must be adjusted to new technologies that will be developed and introduced.

To demonstrate the application of superstructure-based optimization in biofuel supply chains, a biorefinery superstructure model for biodiesel production from microalgae is presented in the following sections. The superstructure model determines the optimal production pathways while taking into account uncertainties related to technical factors.

8.2.1 Processing rout selection for microalgae biorefinery: a case study

The production chain of microalgae biodiesel starts with the cultivation of microalgae in an aqueous system fed continuously with water, CO_2, and nutrients. After the growing cycle is completed, algae are separated from the culture medium in the harvesting step and then go through the drying step to be protected against decay by removing excess water. In the extraction step, lipids are extracted from algae cells and then proceed to the conversion step where the lipids are converted into biodiesel (Mohseni, Pishvaee, & Sahebi, 2016).

Following the model developed by Rizwan, Lee, and Gani (2015), we construct the superstructure of a microalgae-based biorefinery that encompasses available technological options for the different steps performing the operations within the biorefinery. As illustrated in Fig. 8.2, it involves five main steps: (1) microalgae cultivation, (2) harvesting and dewatering the microalgae slurry, (3) drying the dewatered algae for further density increase, (4) lipid extraction, and (5) converting lipids into biodiesel and subsequent purification. Each processing step has several technological options labeled by two indices; the first refers to the option number and the second to its corresponding processing step. An empty box is added in the drying step to allow the bypassing of this step if the wet lipid extraction option is chosen since no drying is required for this extraction method. Moreover, it is assumed that after the harvesting, a fraction of water and nutrients is recycled back to the cultivation step to enhance process efficiency (Mohseni & Pishvaee, 2016).

Given the constructed superstructure, the goal is to find the optimal processing route from microalgae cultivation to biodiesel production while simultaneously determining the optimal amount of flows entering and

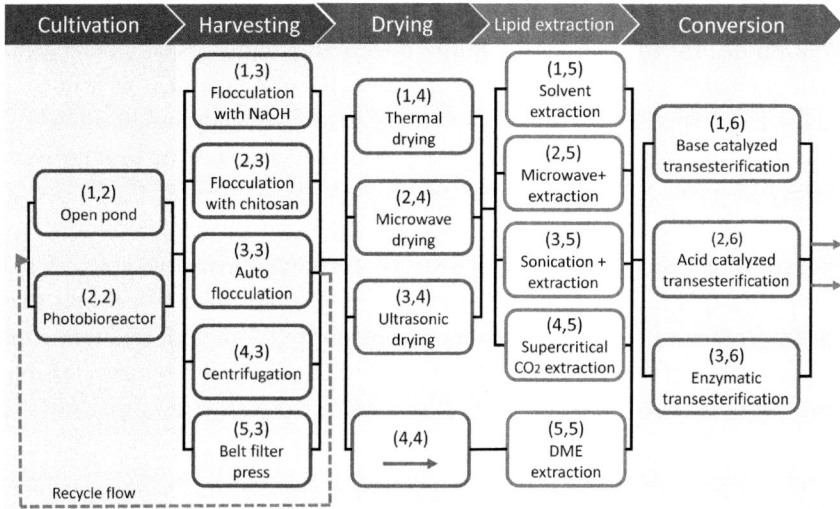

Figure 8.2 *The superstructure of a microalgae-based biorefinery.*

leaving each functional unit of the processing route, resulting in a biorefinery with maximum economic potential.

The performance of the biorefinery superstructure is influenced by uncertainties that arise from the limited knowledge about the production processes and inconsistencies in technical information reported in the literature. These uncertainties must be taken into account by the superstructure optimization model so that the reliability of the optimal biorefinery configuration and its related decisions obtained by the model is ensured in the face of the uncertainties. The results of our sensitivity analysis based on a deterministic version of the model show that varying the percent yield of reactions occurring in the processing units has a significant impact on both the design and profitability of the microalgae-based biorefinery. The accurate estimation of the percent yield, which is defined as the ratio of the actual yield to the theoretical yield of a reaction, is difficult, particularly in the case of biofuel production from microalgae because there is at present no commercial-scale microalgae biorefinery.

Among optimization approaches that can cope with uncertainty, robust scenario-based stochastic programming seems to be a suitable option to deal with uncertainty in the yield parameters due to two main reasons. First, although available information about the yield parameters is inconsistent, it is possible to obtain a set of potential scenarios for them. In such a circumstance, the probability distributions governing the uncertain parameters are difficult to estimate, precluding the use of stochastic programming with continuous distributions. Second, the uncertain parameters of yield appear in an equality constraint (see Eq. (8.5)), which can be handled well by scenario-based approaches, while most fuzzy mathematical programming and robust optimization approaches have a poor performance in dealing with uncertainty in equality constraints.

Robust scenario-based stochastic programming, first introduced by Mulvey, Vanderbei, and Zenios (1995), incorporates two measures of robustness: solution robustness and feasibility robustness. The first means that the optimal solution will not change significantly among different scenarios realizations, and the second means that the optimal solution remains feasible for almost all scenario realizations.

8.2.1.1 Mathematical model

The superstructure problem defined in the previous section is formulated as a mixed-integer linear programming (MILP) model. A list of indices, parameters, and decision variables required to formulate the model is given in Table 8.1.

8.2.1.1.1 Objective function

The objective function of the proposed superstructure model is to minimize the expected biorefinery cost subject to meeting the underlying constraints. It consists of three parts. The first is the annualized investment cost of processing units, the second is the cost of utilities, the third is the cost of materials used in the form of makeup flows, and the last one is the revenue from selling various products (biodiesel and glycerin) leaving the conversion step. The objective function under scenario s is formulated as follows:

$$\min L^s = \sum_{o \in O}\sum_{j \in J} CI_{o,j}Y_{o,j} + \sum_{v \in V}\sum_{o \in O}\sum_{j \in J} CV_v U^s_{v,o,j} + \sum_{i \in I}\sum_{o \in O}\sum_{j \in J} CM_i D^s_{i,o,j} - \sum_{i \in I}\sum_{o \in O} P_i \overline{F}^s_{i,o,6}$$

$$(8.1)$$

Adopting the robust optimization approached developed by Leung, Tsang, Ng and Wu, 2007) (described in detail in Chapter 6), the robust counterpart of the objective function is constructed as follows:

$$\min \sum_{s \in S} \mathrm{pr}^s L^s + \pi \sum_{s \in S} \mathrm{pr}^s \left[\left(L^s - \sum_{s' \in S} \mathrm{pr}^{s'} L^{s'} \right) + 2\theta^s \right] + \sigma \sum_{s \in S} \mathrm{pr}^s \delta^s$$

$$\text{s.t.} \quad L^s - \sum_{s' \in S} \mathrm{pr}^{s'} L^{s'} + \theta^s \geq 0, \quad \forall s \in S \qquad (8.2)$$

Table 8.1 Indices, parameters, and decision variables used in the superstructure model.

Sets

J	Set of processing steps
O	Set of technological options
I	Set of all components present in the superstructure (water, lipid, biodiesel, etc.)
V	Set of utility types
S	Set of scenarios

Subsets

O_j	Technological options belonging to step j
$I_{o,j}$	Components participating in the reaction in option o of step j

Singleton subsets

$R_{o,j}$	The reaction occurring in option o of step j
$M_{o,j}$	Component participating as the main reactant in the reaction in option o of step j
Ik_j	The main component leaving step j

Parameters

$\alpha_{i,r}^s$	Percent ratio of actual yield of component i to its theoretical yield in reaction r under scenario s
$\beta_{i,m}$	Stoichiometric ratio of component i to component m
W_i	Molecular weight of component i
$\mu_{i,o,j}$	Recyclable fraction of component i leaving option o of step j
$\eta_{i,o,j}$	Waste fraction of component i leaving option o of step j
$\omega_{i,o,j}$	Amount of component i added to option o of step j to produce one unit of the main component leaving step j
$\lambda_{v,o,j}$	Amount of utility v added to option o of step j to produce one unit of the main component leaving step j
$CI_{o,j}$	Annualized investment cost of option o of step j
CV_v	Cost of utility v
CM_i	Cost of component i used in the form of makeup flow
P_i	Selling price of product i
pr^s	Probability of occurrence of scenario s
σ	Penalty cost per unit of biodiesel shortage
BT	Biodiesel production target

Positive continuous variables

$FI_{i,o,j}^s$	Inflow of component i entering option o of step j under scenario s
$FO_{i,o,j}^s$	Outflow of component i leaving option o of step j under scenario s
$\overline{F}_{i,o,j}^s$	Flow of component i going from step j to option o of step $j+1$ under scenario s

(Continued)

Table 8.1 Indices, parameters, and decision variables used in the superstructure model. (*Cont.*)

$D^s_{i,o,j}$	Makeup flow of component i entering option o of step j under scenario s
$E^s_{i,o,j}$	Waste flow of component i leaving option o of step j under scenario s
$H^s_{i,o,o',j}$	Recycle flow of component i returning from option o of step j to option o' of step $j-1$ under scenario s
$U^s_{v,o,j}$	Flow of utility v added to option o of step j under scenario s

Binary variables

$Y_{o,j}$	Equal to 1 if technological option o from processing step j is selected; 0 otherwise

where the first term is the expected (mean) value of objective function (8.1) over all scenarios, controlling the average performance of the biorefinery. The second term controls solution robustness by minimizing the deviation of the objective function values from the expected value under different scenarios, and the third term controls feasibility robustness by penalizing the violation allowed in constraint (8.13). σ and π are weighting parameters adjusted by the decision maker to make a trade-off between the three defined terms.

8.2.1.1.2 Constraints

The general flow diagram for a technological option within a processing step is depicted in Fig. 8.3. To formulate mass balance constraints, it is assumed that there is a sequence of three stages in each box: inlet converging,

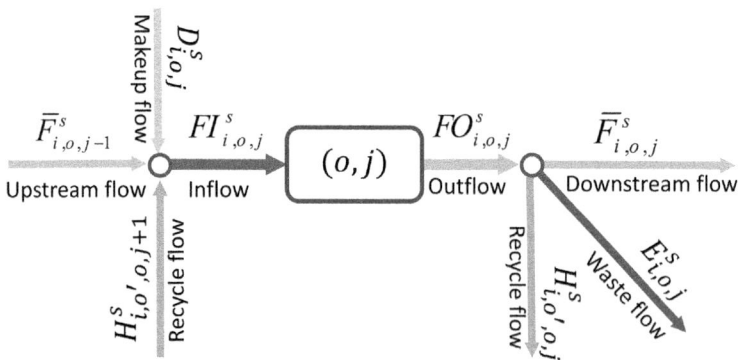

Figure 8.3 *The general flow diagram for a technological option within a processing step.*

conversion, and outlet separation. The first stage taking place at the inlet of the box is modeled by Eq. (8.3), which states that the inflow of component i to option o of processing step j must be equal to the sum of three flows: (1) the upstream flow coming from the previous step, (2) the makeup flow coming from external sources in the form of water, nutrients, solvents, and other chemicals, and (3) recycle flow coming from the next step. It should be noted that in this and all subsequent equations the index j is greater than or equal to 2, and the value of $\overline{F}_{i,o,1}^s$ is set to zero since there is no upstream flow for the cultivation step.

$$\overline{F}_{i,o,j-1}^s + D_{i,o,j}^s + \sum_{o' \in O_{j+1}} H_{i,o',o,j+1}^s = FI_{i,o,j}^s \quad \forall o \in O_j, j \in J,$$

$$i \in I, s \in S \quad (8.3)$$

In the conversion stage, depending on whether or not component i takes part in reaction m that may occur in option o of step j, the outflow is calculated by Eq. (8.4) or (8.5). If component i is not involved in the reaction, its inflow and outflow are equal, as enforced by Eq. (8.4).

$$FI_{i,o,j}^s = FO_{i,o,j}^s \quad \forall o \in O_j, j \in J, i \notin I_{o,j}, s \in S \quad (8.4)$$

Otherwise, the outflow of component i is calculated by the following equation, using the reaction stoichiometry to determine the amount of reactants and products that are consumed or produced in the reaction:

$$FO_{i,o,j}^s = FI_{i,o,j}^s + \left(\tilde{\alpha}_{i,r}^s \beta_{i,m} \frac{FI_{m,o,j}^s}{W_m} \right) W_i \quad \forall o \in O_j, j \in J, r \in R_{o,j},$$

$$m \in M_{o,j}, i \in I_{o,j}, s \in S \quad (8.5)$$

where a new index, m, is introduced to denote the main reactant (i.e., component m), based on which the consumption rate of other reactants and production rate of products are measured. $\beta_{i,m}$ is the stoichiometric ratio of component i to the main reactant, whose value is positive or negative when component i is a product or a reactant, respectively. The uncertain param eter $\tilde{\alpha}_{i,r}^s$ is the percent yield defined as the ratio of the amount of product i actually obtained from reaction r to its stoichiometrically calculated theoretical amount. The value of this parameter is set to one for those components that are employed as reactants and less than one for products. W_m and W_i represent the molecular weight of the main reactant and component i, respectively, and $FI_{m,o,j}^s$ is the incoming flow of the main reactant.

The separation happens at the outlet of the box where the outflow of component i is split into three streams: (1) the downstream flow going to the next processing step, (2) the recycle flow returning to the previous step, and (3) the waste flow sent for disposal, as modeled by Eq. (8.6):

$$\sum_{o \in O_{j+1}} \bar{F}^{s}_{i,o,j} = \sum_{o \in O_{j}} \left(FO^{s}_{i,o,j} - E^{s}_{i,o,j} - H^{s}_{i,o,j} \right) \quad \forall j \in J, \quad i \in I, s \in S \qquad (8.6)$$

Constraint (8.7) guarantees that only one technological option can be selected for each processing step. Moreover, the inflow of component i will be allowed to enter a technological option only if it is selected, which is expressed by constraint (8.8) where M is a sufficiently large positive number to ensure the constraint is redundant when that option is chosen.

$$\sum_{o \in O_{j}} Y_{o,j} \leq 1 \quad \forall j \in J \qquad (8.7)$$

$$FI^{s}_{i,o,j} \leq Y_{o,j} M \quad \forall o \in O_{j}, j \in J, i \in I, s \in S \qquad (8.8)$$

The amount of component i recycled back from step j to step $j - 1$ is determined by Eq. (8.9) where the outflow of component i leaving option o is multiplied by its corresponding recyclable fraction. Similarly, the amount of component i that is not recycled and must be disposed of is calculated by Eq. (8.10)

$$E^{s}_{i,o,j} = \mu_{i,o,j} FO^{s}_{i,o,j} \quad \forall o \in O_{j}, j \in J, i \in I, s \in S \qquad (8.9)$$

$$\sum_{o' \in O_{j-1}} H^{s}_{i,o,o',j} = \eta_{i,o,j} FO^{s}_{i,o,j} \quad \forall o \in O_{j}, j \in J, \quad i \in I, s \in S \qquad (8.10)$$

The makeup flow of component i fed to option o of step j is calculated by Eq. (8.11) where the outflow of the main component leaving that step (indexed as k) is multiplied by the amount of component i that must be added per unit of the main component. Microalgae biomass is considered as the main component leaving the cultivation, harvesting, and drying steps, microalgae lipid is considered for the lipid extraction step, and biodiesel for the conversion step. Following a similar manner to the makeup flow, the amount of different utilities (i.e., electricity and heating energy) consumed by option o of step j is calculated by Eq. (8.12):

$$D_{i,o,j}^{s} = \omega_{i,o,j} FO_{k,o,j}^{s} \quad \forall o \in O_{j}, j \in J, i \in I,$$
$$k \in IK_{j}, s \in S \qquad (8.11)$$

$$U_{v,o,j}^{s} = \lambda_{v,o,j} FI_{k,o,j}^{s} \quad \forall o \in O_{j}, j \in J, v \in V,$$
$$k \in IK_{j}, s \in S \qquad (8.12)$$

Constraint (8.13) indicates that the biodiesel production target is reached by the amount of component 19 (i.e., biodiesel) produced by the conversion step. Incorporating the biodiesel shortage variable δ^s allows the violation of this constraint under some worst-case scenarios, but such violations are penalized in the objective function.

$$\sum_{o \in O} \overline{F}_{19,o,6}^{s} + \delta^{s} = TB \quad \forall s \in S \qquad (8.13)$$

8.2.1.2 Case study

Based on a microalgae-to-biodiesel supply chain design model developed for biodiesel production in the Midwestern states of the U.S. comprising Indiana, Illinois, Kansas, Kentucky, Iowa, Missouri, and Nebraska, the optimal design suggests establishing five biorefineries, each with an annual capacity of 5 million gallons (Nodooshan et al., 2018). In this case study, the proposed model is used to determine the optimal design and synthesis of a microalgae-based biorefinery capable of producing 5 million gallons biodiesel per year. The values of the input parameters such as the stoichiometric ratios, recyclable and waste fractions, molecular weights, investment costs, cost factors of utilities, and other components are taken from Rizwan, Zaman, Lee, and Gani (2015) and Delrue et al. (2012). The uncertainty in the percent yield of each reaction is modeled by three different scenarios (optimistic, moderate, and pessimistic) to cover its possible values reported in the literature (Rizwan, Zaman, et al., 2015) and it is assumed that all scenarios have the same probability to happen.

8.2.1.2.1 Results and discussion

This section evaluates the performance of the proposed robust model and compares it with that of a deterministic superstructure optimization model which is formulated by fixing the uncertainty parameters in the robust model at their nominal values (i.e., the moderate scenario). All computational experiments are conducted by CPLEX 12.5 solver of GAMS software on a desktop PC equipped with an Intel Core i7-950 processor operating at 3.40 GHz.

8.2.1.2.2 Economic performance of the biorefinery

A comparison between the total costs of the biorefinery obtained by the robust and deterministic models is provided in Fig. 8.4. The total cost is $378.5 million in the deterministic model, which is at least 16% lower than that of the robust model. The cost difference between the two models that can be interpreted as "the cost of robustness" increases by increasing the value of the adjustable parameters π and σ, which control the solution and feasibility robustness, respectively. When the value of π is set to 0, the robust model is reduced to a simple scenario-based stochastic model where there is no control on the deviation of the objective function from its expected value under different scenarios. By assigning positive values to π, the robust model minimizes the deviation in addition to the expected total cost, meaning the average economic performance may be somewhat sacrificed to avoid high variations in the total cost that can be caused by different scenarios. The larger the value of π is, the more robust the solution would be; however, this comes at a higher robustness cost. Therefore, varying the value of this parameter enables the decision maker to control the trade-off between the solution robustness and its cost.

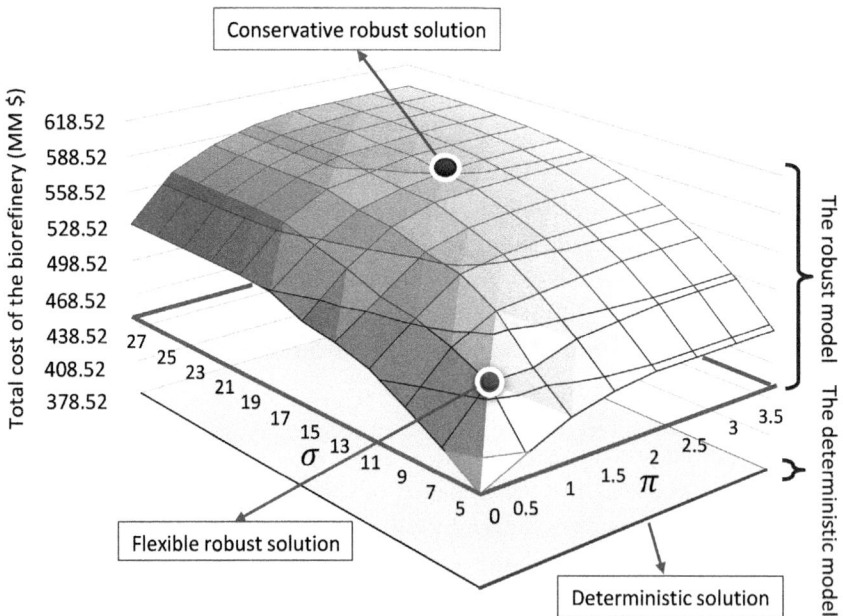

Figure 8.4 *Total costs of the biorefinery obtained by the robust and deterministic models.*

By increasing the value of σ, there is a tendency toward a robust solution that remains feasible for a larger number of scenarios but at the expense of an increase in the total cost. A conservative decision-maker selects a sufficiently large value of σ to avoid any violation of constraint (8.13) (i.e., biodiesel shortage) and must pay the cost of this protection against infeasibility. On the other hand, a risk-taking decision-maker who can tolerate the constraint violation for some scenarios chooses lower values of σ to make a trade-off between the economic and conservative performance.

8.2.1.2.3 Optimal processing pathway for the biorefinery

To allow for different risk preferences in evaluating optimal processing pathways, the following three solutions are investigated:

1. **Deterministic solution.** This solution is obtained from the deterministic version of the robust model, which is formulated by substituting the uncertain parameters with their nominal values.
2. **Conservative robust solution.** As represented in Fig. 8.4, the total cost is insensitive to further increase in the parameters σ and π after they take values of 19 and 2, respectively. These values are selected for the robust model to provide a conservative solution with the highest reliability level in terms of solution and feasibility robustness, although it imposes a significant cost burden on the biorefinery.
3. **Flexible robust solution.** To simultaneously consider the reliability and profitability of the biorefinery, a flexible robust solution is generated by setting the values of σ and π to 7 and 0.5, respectively. The robust superstructure model with these values allows flexibility in controlling solution and feasibility robustness to avoid a severe deterioration in the total cost.

As illustrated in Fig. 8.5, the optimal processing route corresponding to the deterministic solution is composed of microalgae cultivation in open ponds, harvesting by flocculation using chitosan as a flocculent, lipid extraction from wet biomass by dimethyl ether (DME), and acid-catalyzed transesterification producing biodiesel and glycerin as final products. Although this pathway is less costly than the other two pathways that capture the uncertainty, it is highly unreliable as the solution of the deterministic model becomes suboptimal or even infeasible if the actually realized values of the input parameters differ from the nominal values.

The optimal processing route is determined by solving the flexible robust version is given in Fig. 8.5, which consists of open pond cultivation, harvesting by chitosan flocculation, thermal drying, lipid extraction by

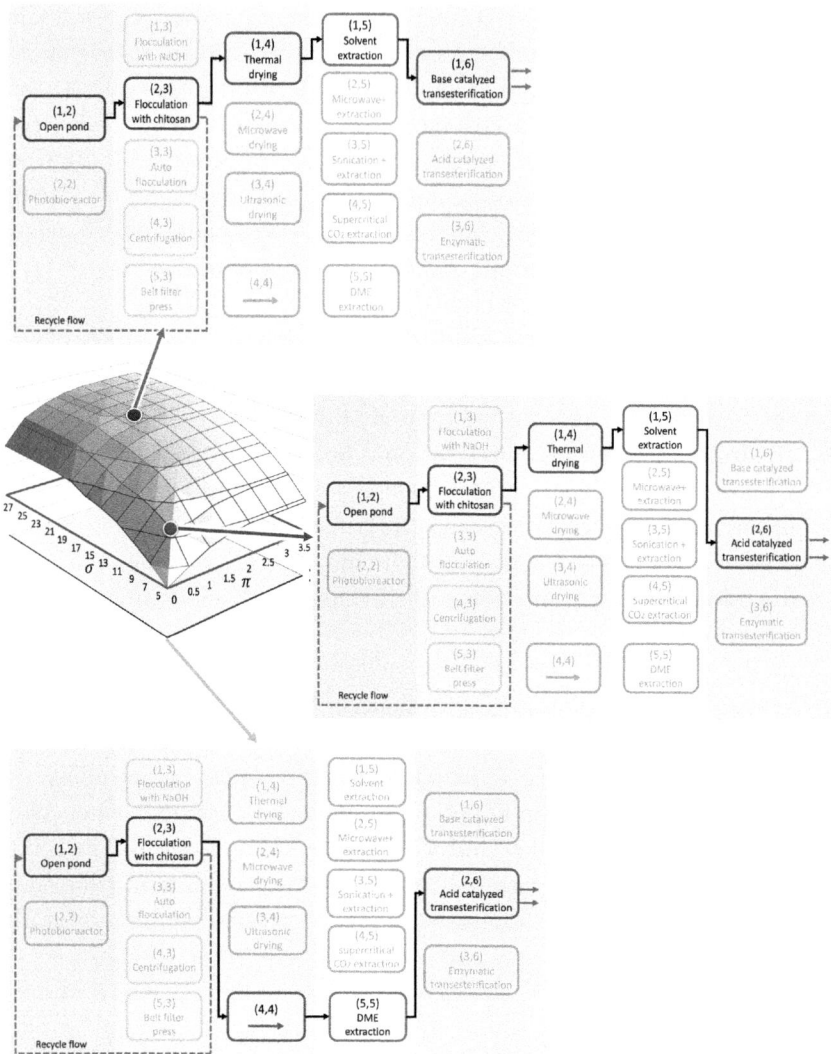

Figure 8.5 *Optimal processing routes obtained by the robust and deterministic models.*

hexane, and acid-catalyzed transesterification. In contrast to the deterministic pathway, here the harvested biomass undergoes thermal drying before
hexane-based lipid extraction rather than being sent directly to the wet
lipid extraction stage, while all the other functional units remain the same.
This implies that adding the energy and cost intensive process of drying that
helps improve lipid yield results in a more robust processing pathway. The
cost of robustness for this pathway is not very high as the flexible robust

solution has a 34% higher total cost compared to the deterministic solution ($507.2 million vs $378.5 million).

As shown in Fig. 8.5, the conservative robust solution selects open pond cultivation, chitosan-based flocculation, thermal drying, hexane extraction, and base–catalyzed transesterification. These processing steps, except for the last one, are the same for both the flexible and conservative robust solutions. In this processing pathway, base–catalyzed transesterification is determined as the optimal conversion method, despite its lower yield compared to acid–catalyzed transesterification. This is because base–catalyzed transesterification is the most commonly used method for biodiesel production in industrial scale and its reaction yield can be estimated with great accuracy; thus, the optimistic and pessimistic values of its yield are very close to each other, making it a more reliable technological option. The cost of the conservative pathway providing maximum solution and feasibility robustness is found to be $588.8 million, approximately 16% and 56% higher than that of the flexible robust and deterministic solutions, respectively. It should be emphasized again that among the deterministic and different robust solutions, the best one is chosen on the basis of the available budget for coping with uncertainty and the risk preferences of the decision maker.

8.3 Biofuel supply chain master planning

As depicted in Fig. 8.6, the planning matrix of biomass supply chains consists of four supply chain modules: biomass supply and pretreatment, biofuel production, biofuel blending and distribution, biofuel sales, which are structured according to the dimension of the planning horizon that encompasses

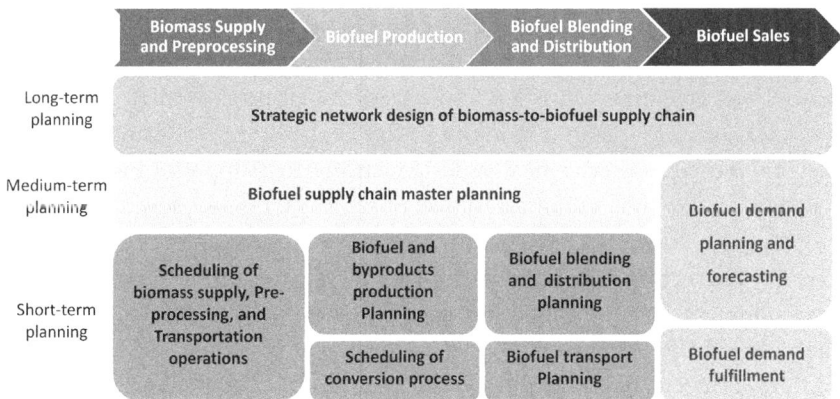

Figure 8.6 *The planning matrix for biofuel supply chains.*

long-term, medium-term, and short-term planning levels (Meyr, Wagner, & Rohde, 2015). The long-term planning level supports decisions on the structure of the biomass supply chain, such as location, capacity, and technology of biorefineries, while the short-term planning level focuses on daily operations such as biomass harvest scheduling.

As an intermediary between strategic and operational planning levels, biofuel supply chain master planning aims to coordinate medium-term decisions related to the supply and pretreatment of biomass, the production of biofuel, and the transportation and storage of biomass and biofuel, such as the quantities of biomass cultivated, harvested, preprocessed, the quantity of biofuel produced, the quantity of biomass and biofuel stored, and the quantity of biomass and biofuel transported from supply areas to biorefineries and from biorefineries to demand zones, respectively. Traditionally, these decisions are made either independently or sequentially, leading to the poor overall performance of the supply chain. A supply chain master plan is generated centrally and adjusted periodically to synchronize the flow of biomass and biofuel along the chain. This enables the supply chain entities (i.e., biomass suppliers, storage units, and biorefineries) to reduce their inventory levels, while without coordination between the entities, larger buffers are required to mitigate the increased risks of stock-outs and ensure a continuous flow of material. In other words, the master planning is responsible for generating an aggregated production and distribution plan for all entities of the supply chain based on the demand data received from the demand planning module (Albrecht, Rohde, & Wagner, 2015; Meyr et al., 2015).

Obtaining a reliable and cost-effective master plan, however, may prove to be difficult in biomass supply chains due to the various sources of uncertainties in different stages. On the supply side, biomass supply is highly influenced by various factors such as climate and weather conditions (e.g., sunlight, pest threats, and precipitation), farmers planting decisions, and natural catastrophic events (e.g., floods, earthquakes, and hurricanes). The limited availability, geographical dispersion, and the seasonal nature of biomass sources are other critical issues governing the supply of biomass. On the production side, technical and economic parameters associated with biofuel production are tainted with uncertainty since most of the technologies for converting biomass into biofuel are still under development, and the commercial-scale production from biomass has not yet achieved. On the demand side, the demand and price of biofuel are highly dependent on various factors such as worldwide variations in fossil fuels and crude oil prices, changing policies and regulations, and public acceptance of biofuels, which makes it difficult to forecast biofuel demand. Therefore, for efficient management and planning of biomass-to-biofuel supply chains, it is crucial to incorporate various sources of uncertainties in the optimization models of supply chain master planning to reliably meet production targets at minimum cost. To illustrate how to provide an optimal planning model for existing biofuel supply chains, a mathematical optimization model for the master planning of the *Jatropha curcas* L. (JCL)-to-biodiesel supply chain under multiple uncertainties is presented in the following.

8.3.1 JCL-to-biodiesel supply chain master planning under uncertainty

The structure of the JCL-based biodiesel supply chain is depicted in Fig. 8.7. JCL seeds are harvested from cultivation sites and shipped to collection and pretreatment centers. At collection and pretreatment centers (1) seeds are separated from the husk, (2) the kernels are separated from the shell, and (3) kernels are cleaned and desiccated before the oil-extraction process. Then, JCL kernels are transported to oil extraction centers, and their oil is extracted via cold-pressing technology and then purified. The byproducts of pretreatment and oil extraction centers are shells/husk, as well as seedcakes that can be sold as fertilizer. At biorefineries, the purified oil is converted to biodiesel (i.e., main product) and glycerin (i.e., byproduct) through transesterification that is the most common process for biodiesel production. Produced biodiesel and glycerin are transported to distribution centers and glycerin consumption centers, respectively. Biodiesel is transported from distribution centers to demand zones and blended there based on a predetermined percentage. It should be noted that all the facilities considered in the concerned biodiesel supply chain are linked via two transportation modes, namely rail, and road. Moreover, trailer trucks and tanker trucks are used to convey solid and liquid materials, respectively.

Compared to traditional supply chains, biofuel supply chains are more vulnerable to risks due to the high level of uncertainties, mainly in biomass supply, biofuel market demand and price, and conversion technologies. These uncertainties manifest themselves as uncertain parameters in the supply chain optimization models that could impact the overall performance and design of the supply chain. Therefore, it is essential to incorporate these uncertainties in supply, production and demand side data when developing optimization models for the planning of biofuel supply chains. As elaborated in the literature, the various sources of uncertainty faced by JCL-based biodiesel supply chain (Babazadeh, Razmi, Pishvaee, & Rabbani, 2017;

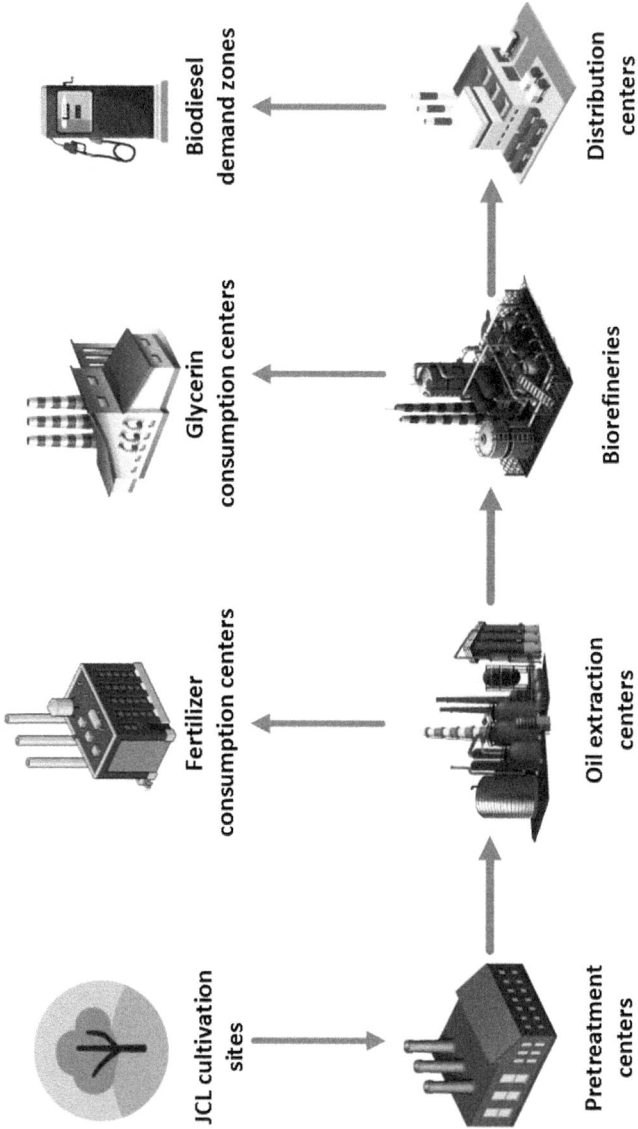

Figure 8.7 *The structure of the JCL-to-biodiesel supply chain.*

Ghelichi, Saidi-Mehrabad, & Pishvaee, 2018) can be categorized into (1) uncertainty of JCL trees yield that is highly dependent on weather conditions, especially amount of rainfall, (2) uncertainty in biodiesel demand and price, as they are sensitive to various factors such as changing regulations and policies, public acceptance of biofuels, crude oil price, increasing competitive pressure and market complexities, and (3) uncertainty in the fixed and variable costs of facilities mainly arising from the fact that the biofuel industry is still in its infancy and large-scale production of biofuels has not been established in a commercial scale, which can highly influence both strategic and tactical level decisions of biodiesel supply chains.

To hedge against these uncertainties, the robust fuzzy programming is adopted in this study since the commercial-scale production from JCL has not yet achieved, and therefore, there is no sufficient historical data to fit a probability distribution for uncertain parameters that is required for applying the continuous stochastic programming approach. On the other hand, the scenario-based stochastic approach may lead to the high computational complexity of the problem because of the huge number of scenarios needed to describe the possible values of uncertain parameters. A reasonable option in these cases may be to fit a possibility distribution for each uncertain parameter according to some limited historical data, as well as subjective knowledge of decision makers (DMs). Therefore, fuzzy numbers can be used to formulate the epistemic uncertainty arising from the lack of complete knowledge about the parameters, and fuzzy programming approaches are the most appropriate tools to deal with such uncertainties. In this regard, a robust fuzzy programming approach, which takes advantage of both fuzzy programming and robust optimization approaches, is applied in this study for coping with a wide range of uncertainties that are faced by biodiesel supply chains.

8.3.1.1 *Mathematical model*

A multiperiod MILP model is developed here to address the JCL-to-biodiesel supply chain master planning problem addressed in the previous section. The proposed mathematical model determines optimal values of tactical decisions in each time period, such as:

- Amount of harvested JCL that is sent to the pretreatment centers,
- Amount of JCL seeds that are sent from pretreatment centers to oil extraction centers,
- Amount of produced biodiesel and glycerin at biorefineries,
- Amount of biodiesel that is sent to the distribution centers,

- Inventory level of JCL at pretreatment centers and oil extraction centers,
- Inventory level of biodiesel and glycerin at biorefineries,
- Selection of transportation mode,
- Amount of capacity expansion at different facilities, including collection and pretreatment centers, oil extraction centers, biorefineries, and distribution centers.

Moreover, JCL yield, costs and prices, and biodiesel demand are assumed to be uncertain, which are modeled as fuzzy numbers. All indices, parameters, and decision variables of the model are introduced in Table 8.2. Note that uncertain parameters are distinguished by a tilde. The objective function and constraints of the proposed model are described in the following.

8.3.1.1.1 Objective function

The objective function maximizes the total profit of the JCL-based biodiesel supply chain, which is equal to total revenue minus total costs of the chain. Verbal representation of the objective function and its components are expressed by Eqs. (8.14)–(8.20):

$$\max \ Z = \text{Total revenue from selling biodiesel, glycerin, and fertilizers} - \text{Production cost} - \text{Inventory holding cost} - \text{Capacity expansion cost} - \text{diesel purchasing cost} \tag{8.14}$$

$$\text{Total revenue} = \sum_{b,n,m,t} \widetilde{pr}_t^{bl} f_{bnmt}^{bld} + \sum_{k,f,m,t} \widetilde{pr}_t^{fz} f_{kfmt}^{fzo} + \sum_{j,f,m,t} \widetilde{pr}_t^{fz} f_{jfmt}^{fzp} + \sum_{l,g,m,t} \widetilde{pr}_t^{gl} f_{lgmt}^{gl} \tag{8.15}$$

$$\text{Production cost} = \sum_{i,t} \widetilde{cul}_{it} la_{i,t} + \sum_{i,t} \widetilde{upc}_{it} pj_{it} + \sum_{j,t} \widetilde{uptc}_{jt} ps_{jt} + \sum_{k,t} \widetilde{uec}_{kt} po_{kt} + \sum_{l,t} \widetilde{bpc}_{lt} pb_{lt} + \sum_{l,t} \widetilde{gpc}_{lt} pg_{lt} \tag{8.16}$$

$$\text{Inventory holding cost} = \sum_{j,t} \widetilde{jh}_{jt} ij_{jt} + \sum_{k,t} \widetilde{oh}_{kt} io_{kt} + \sum_{l,t} \widetilde{bhr}_{lt} ib_{lt} + \sum_{l,t} \widetilde{gh}_{lt} ig_{lt} + \sum_{b,t} \widetilde{bhd}_{bt} ibd_{bt} \tag{8.17}$$

$$\text{Transportation cost} = \sum_{i,j,m,t} \tilde{c}_{ijmt}^{jy} f_{ijmt}^{jy} + \sum_{j,k,m,t} \tilde{c}_{jkmt}^{js} f_{jkmt}^{js} + \sum_{k,l,m,t} \tilde{c}_{klmt}^{jo} f_{klmt}^{jo} + \sum_{l,b,m,t} \tilde{c}_{lbmt}^{blr} f_{lbmt}^{blr} + \sum_{b,n,m,t} \tilde{c}_{bnmt}^{bld} f_{bnmt}^{bld} + \sum_{l,g,m,t} \tilde{c}_{lgmt}^{gl} f_{lgmt}^{gl} + \sum_{k,f,m,t} \tilde{c}_{kfmt}^{fzo} f_{kfmt}^{fzo} + \sum_{j,f,m,t} \tilde{c}_{jfmt}^{fzp} f_{jfmt}^{fzp} \tag{8.18}$$

Table 8.2 Indices, parameters, and decision variables used in the biodiesel supply chain master planning model.

Indices

i	Index of JCL cultivation sites; $i = (1,\ldots,I)$
j	Index of JCL collection and pretreatment centers; $j = (1,\ldots,J)$
k	Index of oil extraction centers; $k = (1,\ldots,K)$
l	Index of biorefineries; $l = (1,\ldots,L)$
b	Index of biodiesel distribution centers; $b = (1,\ldots,B)$
n	Index of consumer centers of biodiesel; $n = (1,\ldots,N)$
g	Index of consumer centers of glycerin; $g = (1,\ldots,G)$
f	Index of consumer centers of fertilizers; $f = (1,\ldots,F)$
m	Index of transportation modes (road and railway); $m = (1,\ldots,M)$
t	Index of time periods; $t = (1,\ldots,T)$

Parameters

la_{it}	Allocated land area of cultivation site i for cultivating JCL in time period t (ha/period)
$\widetilde{upc}_{it}$	Unit harvesting and collection cost of JCL yields from cultivation site i in time period t ($\$.ha^{-1}$/period)
$\widetilde{cul}_{it}$	Unit cultivation cost of JCL in cultivation site i at time period t ($\$.ha^{-1}$/period)
$\widetilde{uptc}_{jt}$	Unit pretreatment cost of JCL seeds at collection and pretreatment center j in time period t ($\$.ton^{-1}$/period)
$\widetilde{uec}_{kt}$	Unit oil extraction cost from JCL yields at oil extraction center k in time period t ($\$.ton^{-1}$/period)
$\widetilde{bpc}_{lt}$	Unit production cost of biodiesel at biorefinery l in time period t ($\$.ton^{-1}$/period)
$\widetilde{gpc}_{lt}$	Unit production cost of glycerin at biorefinery l in time period t ($\$.ton^{-1}$/period)
$\widetilde{jh}_{jt}$	Unit inventory holding cost of JCL yields at collection and pretreatment center j in time period t ($\$.ton^{-1}$/period)
$\widetilde{oh}_{kt}$	Unit inventory holding cost of JCL oil at collection and pretreatment center k in time period t ($\$.ton^{-1}$/period)
$\widetilde{bhr}_{lt}$	Unit inventory holding cost of biodiesel at biorefinery l in time period t ($\$.ton^{-1}$/period)
$\widetilde{gh}_{lt}$	Unit inventory holding cost of glycerin at biorefinery l in time period t ($\$.ton^{-1}$/period)
$\widetilde{bhd}_{bt}$	Unit inventory holding cost of biodiesel at distribution center b in time period t ($\$.ton^{-1}$/period)
$\tilde{c}_{ijmt}^{jy}$	Transportation cost of JCL yields from cultivation site i to collection and pretreatment center j by transportation mode m in time period t ($\$.ton^{-1}$/period)
$\tilde{c}_{jkmt}^{js}$	Transportation cost of JCL seeds from collection and pretreatment center j to oil extraction center k by transportation mode m in time period t ($\$.ton^{-1}$/period)

(Continued)

Table 8.2 Indices, parameters, and decision variables used in the biodiesel supply chain master planning model. (*Cont.*)

$\tilde{c}_{klmt}^{jo}$	Transportation cost of JCL oil from oil extraction center k to biorefinery l by transportation mode m in time period t (\$.ton^{-1}/period)
$\tilde{c}_{lbmt}^{blr}$	Transportation cost of biodiesel from biorefinery l to distribution center b by transportation mode m in time period t (\$.ton^{-1}/period)
$\tilde{c}_{bnmt}^{bld}$	Transportation cost of biodiesel from distribution center b to demand zone n by transportation mode m in time period t (\$.ton^{-1}/period)
$\tilde{c}_{lg\,mt}^{gl}$	Transportation cost of glycerin from biorefinery l to consumer center g by transportation mode m in time period t (\$.ton^{-1}/period)
$\tilde{c}_{kfmt}^{fzo}$	Transportation cost of fertilizer from oil extraction center k to consumer center f by transportation mode m in time period t (\$.ton^{-1}/period)
$\tilde{c}_{jfmt}^{fzp}$	Transportation cost of fertilizer from collection and pretreatment center j to consumer center f by transportation mode m in time period t (\$.ton^{-1}/period)
$\widetilde{uep}_{jt}$	Unit cost of capacity expansion at collection and pretreatment center j in time period t (\$/ton)
$\widetilde{ueo}_{kt}$	Unit cost of capacity expansion at oil extraction center k in time period t (\$/ton)
$\widetilde{ueb}_{lt}$	Unit cost of capacity expansion at biorefinery l in time period t (\$/ton)
$\widetilde{ued}_{bt}$	Unit cost of capacity expansion at storage and distribution center b in time period t (\$/ton)
$\widetilde{ucd}_{t}$	Unit cost of supplying diesel fuel in time period t (\$/ton)
$\widetilde{pr}_{t}^{bl}$	Selling price of biodiesel in time period t (\$/ton)
$\widetilde{pr}_{t}^{fz}$	Selling price of fertilizer in time period t (\$/ton)
$\widetilde{pr}_{t}^{gl}$	Selling price of glycerin in time period t (\$/ton)
θ	Conversion factor of JCL fruits to JCL seeds (percent)
η	Conversion factor of JCL seeds to oil (percent)
λ	Conversion factor of JCL oil to biodiesel (percent)
$\tilde{d}_{nt}$	Demand of consumer center n in time period t (ton/period)
$\widetilde{yl}_{it}$	Yield of JCL per hectare at cultivation site location i in time period t (ton/ha)

Positive continuous variables

pj_{it}	Amount of harvested JCL from cultivation site i in time period t (ton/period)
ps_{jt}	Amount of pretreated JCL seeds at collection and pretreatment center j in time period t (ton/period)
pf_{jt}	Amount of JCL fertilizer at collection and pretreatment center j in time period t (ton/period)
pof_{kt}	Amount of JCL fertilizer at oil extraction center k j in time period t (ton/period)
po_{kt}	Amount of produced oil at oil extraction center k in time period t (ton/period)

Table 8.2 Indices, parameters, and decision variables used in the biodiesel supply chain master planning model. (*Cont.*)

pb_{lt}	Amount of produced biodiesel at biorefinery l in time period t (ton/period)
pg_{lt}	Amount of produced glycerin at biorefinery l in time period t (ton/period)
iy_{jt}	Inventory level of JCL yields at collection and pretreatment center j in time period t (ton/period)
io_{kt}	Inventory level of JCL oil at oil extraction center k in time period t (ton/period)
ib_{lt}	Inventory level of biodiesel at biorefinery l in time period t (ton/period)
ig_{lt}	Inventory level of glycerin at biorefinery l in time period t (ton/period)
ibd_{bt}	Inventory level of biodiesel at distribution center b in time period t (ton/period)
f^{jy}_{ijmt}	Quantity of JCL yields transported from cultivation site i to collection and pretreatment center j through transportation mode m in time period t (ton/period)
f^{js}_{jkmt}	Quantity of JCL seeds transported from collection and pretreatment center j to oil extraction center k through transportation mode m in time period t (ton/period)
f^{jo}_{klmt}	Quantity of JCL oil transported from oil extraction center k to biorefinery l through transportation mode m in time period t (ton/period)
f^{blr}_{lbmt}	Quantity of biodiesel transported from biorefinery l to distribution center b through transportation mode m in time period t (ton/period)
f^{bld}_{bnmt}	Quantity of biodiesel transported from distribution center b to demand zone n through transportation mode m in time period t (ton/period)
f^{gl}_{lgmt}	Quantity of glycerin transported from biorefinery l to consumer center g through transportation mode m in time period t (ton/period)
f^{fzo}_{kfmt}	Quantity of fertilizer transported from oil extraction center k to consumer center f through transportation mode m in time period t (ton/period)
f^{fzp}_{jfmt}	Quantity of fertilizer transported from collection and pretreatment center j to consumer center f through transportation mode m in time period t (ton/period)
cap^{cp}_{jt}	Capacity of collection and pretreatment center j in time period t (ton/period)
cap^{oc}_{kt}	Capacity of oil extraction center k in time period t (ton/period)
cap^{br}_{lt}	Capacity of biorefinery l in time period t (ton/period)
cap^{dc}_{bt}	Capacity of distribution center b in time period t (ton/period)
ep_{jt}	Amount of capacity expansion at collection and pretreatment center j in time period t (ton/period)
eo_{kt}	Amount of capacity expansion at oil extraction center k in time period t (ton/period)

(*Continued*)

Table 8.2 Indices, parameters, and decision variables used in the biodiesel supply chain master planning model. (*Cont.*)

eb_{lt}	Amount of capacity expansion at biorefinery l in time period t (ton/period)
ed_{bt}	Amount of capacity expansion at distribution center b in time period t (ton/period)
$upexp_j^{cp}$	Maximum amount of capacity that can be added to the existing capacity of collection and pretreatment center j in time period t (ton/period)
$upexp_k^{oc}$	Maximum amount of capacity that can be added to the existing capacity of oil extraction center k in time period t (ton/period)
$upexp_l^{br}$	Maximum amount of capacity that can be added to the existing capacity of biorefinery l in time period t (ton/period)
$upexp_b^{dc}$	Maximum amount of capacity that can be added to the existing capacity of distribution center b in time period t (ton/period)
dc_{nt}	Diesel consumption by consumer n in time period t (ton/period)

$$\text{Capacity expansion cost} = \sum_{j,t} \widetilde{uep}_{jt} \, ep_{jt} + \sum_{k,t} \widetilde{ueo}_{kt} \, eo_{kt} + \sum_{l,t} \widetilde{ueb}_{lt} \, eb_{lt} + \sum_{b,t} \widetilde{ued}_{bt} \, ed_{bt} \tag{8.19}$$

$$\text{Diesel purchasing cost} = \sum_{n,t} \widetilde{ucd}_t \, dc_{nt} \tag{8.20}$$

8.3.1.1.2 Constraints
- Mass balancing

Constraints (8.21)–(8.24) ensure that, at each time period, the harvested JCL yields, JCL seeds, and fertilizers are sent to their corresponding destination facilities.

$$pj_{it} = \sum_{j,m} f_{ijmt}^{jy} \qquad \forall i,t \tag{8.21}$$

$$\sum_{k,m} f_{jkmt}^{jy} = ps_{jt} \qquad \forall j,t \tag{8.22}$$

$$\sum_{f,m} f_{jfmt}^{fzp} = pf_{jt} \qquad \forall j,t \tag{8.23}$$

$$\sum_{f,m} f_{kfmt}^{fzo} = pof_{kt} \qquad \forall k,t \tag{8.24}$$

- Production and conversion constraints

Constraint (8.25) ensures that the amount of JCL that can be harvested from each cultivation site in each time period is not greater than the biomass yield of that cultivation site.

$$pj_{it} \leq \tilde{yl}_{it} la_{i,t} \quad \forall i,t \tag{8.25}$$

Constraints (8.26) and (8.27) determine the amount of JCL yields converted to pretreated seeds and the residue of the pretreatment process that can be used as fertilizers.

$$ps_{jt} = \theta \; iy_{jt} \quad \forall j,t \tag{8.26}$$

$$pf_{jt} = (1-\theta) iy_{jt} \quad \forall j,t \tag{8.27}$$

Constraints (8.28) and (8.29) state that in each oil extraction center and at each time period, the JCL oil is extracted from JCL seeds, and the residue of the extraction process is used as fertilizers.

$$po_{kt} = \eta \sum_{j,m} f^{js}_{jkmt} \quad \forall k,t \tag{8.28}$$

$$pof_{kt} = (1-\eta) \sum_{j,m} f^{js}_{jkmt} \quad \forall k,t \tag{8.29}$$

Constraints (8.30) and (8.31) ensure that in each biorefinery and at each time period, JCL oil is converted to biodiesel and glycerin.

$$pb_{lt} = \lambda \sum_{k,m} f^{jo}_{klmt} \quad \forall l,t \tag{8.30}$$

$$pg_{lt} = (1-\lambda) \sum_{k,m} f^{jo}_{klmt} \quad \forall l,t \tag{8.31}$$

- Inventory balancing

Constraint (8.32) ensures that in each collection and pretreatment center, the amount of JCL yields at each time period is equal to the inventory of JCL yields left from the previous period plus the JCL yields transported from cultivation sites minus the JCL yields converted to JCL seeds. Likewise, constraints (8.33)–(8.36) ensure the inventory balance of JCL oil, glycerin, and biodiesel at their related facilities.

$$ij_{jt} = ij_{j,t-1} + \sum_{i,m} f^{jy}_{ijmt} - \left(\frac{1}{\alpha}\right) ps_{jt} \quad \forall j,t \tag{8.32}$$

$$io_{kt} = io_{k,t-1} + po_{kt} - \sum_{l,m} f_{klmt}^{jo} \quad \forall k,t \tag{8.33}$$

$$ib_{lt} = ib_{l,t-1} + pb_{lt} + \sum_{b,m} f_{lbmt}^{blr} \quad \forall l,t \tag{8.34}$$

$$ig_{lt} = ig_{l,t-1} + pg_{lt} - \sum_{g,m} f_{lgmt}^{gl} \quad \forall l,t \tag{8.35}$$

$$ibd_{bt} = ibd_{b,t-1} + \sum_{l,m} f_{l,b,m}^{blr} - \sum_{n,m} f_{bnmt}^{blr} \quad \forall b,t \tag{8.36}$$

- Capacity expansion and limitations of existing facilities

Constraints (8.37)–(8.44) state that the capacity of (existing) facilities, including collection and pretreatment centers, oil extraction centers, biorefineries, and distribution centers, is dependent on time periods, which can be expanded according to increase of biodiesel demands. It is clear that the capacity of each facility in each time period is equal to the capacity determined in the previous period, and the amount of capacity expanded in the current time period. It should be noted that the capacity of facilities at the beginning of the planning period (i.e., $cap_{j,t=0}^{cp}, cap_{kt}^{oc}, cap_{lt}^{br}, cap_{bt}^{dc}$) is known. On the other hand, constraints (8.45)–(8.48) ensure that the quantity of transported materials to each facility does not exceed its corresponding capacity.

$$cap_{jt}^{cp} = cap_{j,t-1}^{cp} + ep_{jt} \quad \forall j,t \tag{8.37}$$

$$0 \le ep_{jt} \le upexp_j^{cp} \quad \forall j,t \tag{8.38}$$

$$cap_{kt}^{oc} = cap_{k,t-1}^{oc} + eo_{kt} \quad \forall k,t \tag{8.39}$$

$$0 \le eo_{kt} \le upexp_k^{oc} \quad \forall k,t \tag{8.40}$$

$$cap_{lt}^{br} = cap_{l,t-1}^{br} + eb_{lt} \quad \forall l,t \tag{8.41}$$

$$0 \le eb_{lt} \le upexp_l^{br} \quad \forall l,t \tag{8.42}$$

$$cap_{bt}^{dc} = cap_{b,t-1}^{dc} + ed_{bt} \quad \forall b,t \tag{8.43}$$

$$0 \le ed_{bt} \le upexp_b^{dc} \quad \forall b,t \tag{8.44}$$

$$\sum_{i,m} f_{ijmt}^{jy} \le cap_{jt}^{cp} \quad \forall j,t \tag{8.45}$$

$$\sum_{j,m} f_{jkmt}^{js} \le cap_{kt}^{oc} \quad \forall k,t \tag{8.46}$$

$$\sum_{k,m} f^{jo}_{klmt} \leq cap^{br}_{lt} \qquad \forall l,t \qquad (8.47)$$

$$\sum_{l,m} f^{blr}_{lbmt} \leq cap^{dc}_{bt} \qquad \forall b,t \qquad (8.48)$$

- Demand satisfaction

Constraint (8.49) ensures that, in each demand zone and at each time period, the fuel demand is satisfied by the transported biodiesel from distribution centers. Moreover, the unmet demand is satisfied through purchasing diesel as an alternative fuel that can compete with biodiesel in fulfilling demands.

$$\sum_{b,m} f^{bld}_{bnmt} + dc_{nt} \geq \tilde{d}_{nt} \; \forall n,t \qquad (8.49)$$

8.3.1.2 Robust counterpart formulations

In this section, the above deterministic biodiesel supply chain master planning model is extended to a robust counterpart model, taking into account uncertainty in JCL yield, JCL production and preprocessing costs, oil extraction costs, biodiesel production cost, inventory holding cost of materials at different facilities, transportation cost, as well as biodiesel demand and price. To cope with uncertainty, the robust possibilistic programming proposed by Pishvaee, Razmi, and Torabi (2012), which is a powerful modeling approach that takes advantage of both robust optimization theory and fuzzy mathematical programming, is adopted. In the robust fuzzy approach, uncertain parameters are modeled as fuzzy numbers that are characterized by possibility distributions. Possibility distribution, which can be extracted based on some available objective data and subjective opinions of DMs, displays the occurrence degree of all possible values of fuzzy number in the corresponding interval. According to the shape of the possibility distribution, several types of fuzzy numbers can be distinguished, among which triangular and trapezoidal fuzzy numbers are the most common types in practical applications. Here, each uncertain parameter $\tilde{\xi}$ is modeled as a trapezoidal fuzzy number, which is defined by its four prominent points as $\tilde{\xi} = \left(\xi_{(1)}, \xi_{(2)}, \xi_{(3)}, \xi_{(4)} \right)$ (see Fig. 8.8). According to this assumption, the robust counterparts of the constraints and objective function are derived in the following.

8.3.1.2.1 Robust counterparts of the constraints

Following the robust possibilistic approach explained in Chapter 6, constraints (8.25) and (8.49) including the uncertain parameters of

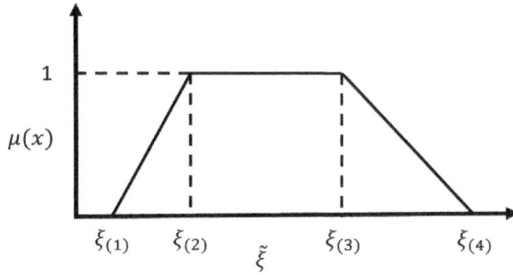

Figure 8.8 *The trapezoidal possibility distribution of the fuzzy parameter $\tilde{\xi}$.*

yield and demand, which are presented as trapezoidal fuzzy numbers, $\left(\text{i.e., } \tilde{yl}_{it} = \left(yl_{it,(1)}, yl_{it,(2)}, yl_{it,(3)}, yl_{it,(4)}\right), \text{ and } \tilde{d}_{nt} = \left(d_{nt,(1)}, d_{nt,(2)}, d_{nt,(3)}, d_{nt,(4)}\right)\right)$, are transformed into the following robust counterparts:

$$pj_{it} \leq \left(\alpha yl_{it,(1)} + (1-\alpha) yl_{it,(2)}\right) la_{i,t} \quad \forall i,t \tag{8.50}$$

$$\sum_{b,m} f_{bnmt}^{bld} + dc_{nt} \geq (1-\beta) d_{nt,(3)} + \beta d_{nt,(4)} \quad \forall n,t \tag{8.51}$$

$$0.5 < \alpha \leq 1 \tag{8.52}$$

where α and β represent the confidence level of corresponding chance constraints.

8.3.1.2.2 Robust counterpart of the objective function
Let us consider the objective function of the above model in the following compact form:

$$\max z = \tilde{p}x_1 - \tilde{q}x_2 \tag{8.53}$$

where x_1 and x_2 denote continuous variables, and p and q correspond to the revenue from selling biodiesel and byproducts, and variable costs of production, transportation, inventory holding, capacity expansion, and diesel purchasing costs, respectively. According to the robust possibilistic approach, the robust counterpart of the above objective function is obtained as follows:

$$\max E[z] - \gamma\left(z_{max} - z_{min}\right) - \delta_1\left[d_{nt,(4)} - (1-\beta)d_{nt,(3)} - \beta d_{nt,(4)}\right]$$
$$- \delta_2\left[\alpha yl_{it,(1)} + (1-\alpha)yl_{it,(2)} - yl_{it,(1)}\right] \tag{8.54}$$

$$E[z] = \left(\frac{p_{(1)} + p_{(2)} + p_{(3)} + p_{(4)}}{4}\right)x_1 - \left(\frac{q_{(1)} + q_{(2)} + q_{(3)} + q_{(4)}}{4}\right)x_2 \tag{8.55}$$

$$z_{max} = p_{(4)}x_1 - q_{(1)}x_2 \tag{8.56}$$

$$z_{min} = p_{(1)}x_1 - q_{(4)}x_2 \tag{8.57}$$

$$x \geq 0, \ 0.5 < \alpha, \beta \leq 1 \tag{8.58}$$

where δ_1 and δ_2 are penalties introduced for possible violation of the chance constraints, respectively, and γ denotes the relative importance of the second term (i.e., the maximum deviation between possible values of z) compared to other terms of the objective function.

8.4 Conclusions

Tactical decisions in the supply chain, which are medium-term decisions that typically span from a few months to one year, identify medium-term policies associated with different processes of the supply chain, according to the network structure established in the strategic-level decision-making. This chapter provides a framework for the tactical decisions that are made at different stages of biomass-to-biofuel supply chains, and discusses uncertainties in biomass supply chains, including biomass yield, biofuel demand and price, and biomass conversion rates. Biofuel supply chain master production planning seeks to provide a production plan to produce biofuels by utilizing the available conversion capacity of biorefineries in a timely and cost-effective manner. To illustrate how to provide an optimal planning model for biofuel supply chains, a multiperiod MILP model is presented in this chapter to address the master planning of the JCL-to-biodiesel supply chain, which determines the optimal values of tactical decisions in each time period. The deterministic biodiesel supply chain master planning model is then extended to a robust counterpart model, taking into account uncertainty in JCL yield, JCL production and preprocessing costs, oil extraction costs, biodiesel production cost, inventory holding cost of materials at different facilities, transportation cost, as well as biodiesel demand and price. The robust possibilistic programming is adopted to deal with uncertain parameters that are modeled as fuzzy numbers.

To address the problem of biorefinery process synthesis and design, a biorefinery superstructure model for biodiesel production from microalgae is proposed, which determines the optimal production pathways while taking into account uncertainties related to technical factors. To cope with uncertainty in the yield parameters of the proposed model, robust scenario-based stochastic programming is applied. The results of the sensitivity analysis based on a deterministic version of the model show that varying the percent yield of reactions occurring in the processing units has a significant impact on both the design and profitability of the microalgae-based biorefinery. However, the accurate estimation of the percent yield, which is defined as the ratio of the actual yield to the theoretical yield of a reaction, is difficult, particularly in the case of biofuel production from microalgae because there is at present no commercial-scale microalgae biorefinery.

References

Albrecht, M., Rohde, J., & Wagner, M. (2015). Master planning. In *Supply chain management and advanced planning*. Springer.

An, H., Wilhelm, W. E., & Searcy, S. W. (2011). Biofuel and petroleum-based fuel supply chain research: A literature review. *Biomass and Bioenergy, 35*, 3763–3774.

Babazadeh, R., Razmi, J., Pishvaee, M. S., & Rabbani, M. (2017). A sustainable second-generation biodiesel supply chain network design problem under risk. *Omega, 66*, 258–277.

Bairamzadeh, S., Saidi-Mehrabad, M., & Pishvaee, M. S. (2018). Modelling different types of uncertainty in biofuel supply network design and planning: A robust optimization approach. *Renewable Energy, 116*, 500–517.

Bao, B., Ng., D. K., Tay, D. H., Jiménez-Gutiérrez, A., & El-Halwagi, M. M. (2011). A shortcut method for the preliminary synthesis of process-technology pathways: An optimization approach and application for the conceptual design of integrated biorefineries. *Computers & Chemical Engineering, 35*, 1374–1383.

Church, R. L., Murray, A. T., & Barber, K. H. (2000). Forest planning at the tactical level. *Annals of Operations Research, 95*, 3–18.

Delrue, F., Setier, P. -A., Sahut, C., Cournac, L., Roubaud, A., Peltier, G., & Froment, A. -K. (2012). An economic, sustainability, and energetic model of biodiesel production from microalgae. *Bioresource Technology, 111*, 191–200.

Ebadian, M., Sokhansanj, S., & Webb, E. (2017). Estimating the required logistical resources to support the development of a sustainable corn stover bioeconomy in the USA. *Biofuels, Bioproducts and Biorefining, 11*, 129–149.

Ghelichi, Z., Saidi-Mehrabad, M., & Pishvaee, M. S. (2018). A stochastic programming approach toward optimal design and planning of an integrated green biodiesel supply chain network under uncertainty: A case study. *Energy, 156*, 661–687.

Malladi, K. T., & Sowlati, T. (2018). Biomass logistics: A review of important features, optimization modeling and the new trends. *Renewable and Sustainable Energy Reviews, 94*, 587–599.

Leung, S. C., Tsang, S. O. Ng, W. L. & Wu, Y. (2007). A robust optimization model for multi-site production planning problem in an uncertain environment. *European journal of operational research, 181*(1), 224–238.

Mencarelli, L., Chen, Q., Pagot, A., & Grossmann, I. E. (2020). A review on superstructure optimization approaches in process system engineering. *Computers & Chemical Engineering, 136*, 106808.

Meyr, H., Wagner, M., & Rohde, J. (2015). Structure of advanced planning systems. In *Supply chain management and advanced planning*. Springer.

Mohseni, S., & Pishvaee, M. S. (2016). A robust programming approach towards design and optimization of microalgae-based biofuel supply chain. *Computers & Industrial Engineering, 100*, 58–71.

Mohseni, S., Pishvaee, M. S., & Sahebi, H. (2016). Robust design and planning of microalgae biomass-to-biodiesel supply chain: A case study in Iran. *Energy*, *111*, 736–755.

Mulvey, J. M., Vanderbei, R. J., & Zenios, S. A. (1995). Robust optimization of large-scale systems. *Operations Research*, *43*, 264–281.

Nodooshan, K. G., Moraga, R. J., Chen, S. -J. G., Nguyen, C., Wang, Z., & Mohseni, S. (2018). Environmental and economic optimization of algal biofuel supply chain with multiple technological pathways. *Industrial & Engineering Chemistry Research*, *57*, 6910–6925.

Pishvaee, M. S., Razmi, J., & Torabi, S. A. (2012). Robust possibilistic programming for socially responsible supply chain network design: A new approach. *Fuzzy sets and Systems*, *206*, 1–20.

Rizwan, M., Lee, J. H., & Gani, R. (2015a). Optimal design of microalgae-based biorefinery: Economics, opportunities and challenges. *Applied Energy*, *150*, 69–79.

Rizwan, M., Zaman, M., Lee, J. H., & Gani, R. (2015b). Optimal processing pathway selection for microalgae-based biorefinery under uncertainty. *Computers & Chemical Engineering*, *82*, 362–373.

Yue, D., & You, F. (2016). Optimal supply chain design and operations under multi-scale uncertainties: Nested stochastic robust optimization modeling framework and solution algorithm. *AIChE Journal*, *62*, 3041–3055.

Yue, D., You, F., & Snyder, S. W. (2014). Biomass-to-bioenergy and biofuel supply chain optimization: Overview, key issues and challenges. *Computers & Chemical Engineering*, *66*, 36–56.

Zandi Atashbar, N., Labadie, N., & Prins, C. (2018). Modelling and optimisation of biomass supply chains: A review. *International Journal of Production Research*, *56*, 3482–3506.

CHAPTER 9

Operational planning in biofuel supply chain under uncertainty

9.1 Introduction

Operational decisions in the supply chain are those decisions that are made weekly, daily, or even hourly based on the type and characteristics of the chain. These decisions focus on executing the short-term activities to ensure the optimal flow of materials along the entire chain and continuous operations of the facilities working in the chain. Unlike the strategic decisions that remain unchanged many years, operational decisions can be altered and revised more frequently with regard to different internal and external factors affecting the performance of the supply chain. Operational decisions in the biomass-to-biofuel supply chain include a wide range of activities such as scheduling of biomass harvest and collection operations, allocating equipment and workers to supply areas for the harvest and collection of biomass, operational logistics planning for transporting biomass and biofuel across the chain, scheduling of processing tasks in biorefineries, inventory control in biomass and biofuel storage units, and daily and weekly forecasting of the final customers' biofuel demand (Balaman, 2019; Yue, You, & Snyder, 2014). Fig. 9.1 provides an overview of the operational decisions involved in the biomass supply chain. These decisions are described in detail in Chapter 3.

Short-term planning and scheduling of harvesting and collection operations, as one of the key operational decisions in the upstream of biomass supply chains, aim to obtain the required biomass for biorefineries from supply areas in a cost-effective and timely manner. For this purpose, a series of sequential harvesting and collection operations (e.g., cutting, raking, baling, gathering, and truck loading) must be performed by different types of vehicles (e.g., balers) in geographically dispersed fields within a specified time window. The collection seasons of agriculture-based and forest-based biomass may overlap due to the fact that most biomass types have similar short harvesting seasons. The overlap of these restricted time frames may lead to competition between biomass suppliers in terms of both equipment and labor. Moreover, the timing and frequency of harvest may impact on the yield of biomass. As a result, it is necessary to schedule the harvest

Biomass to Biofuel Supply Chain Design and Planning under Uncertainty
http://dx.doi.org/10.1016/B978-0-12-820640-9.00009-X
247

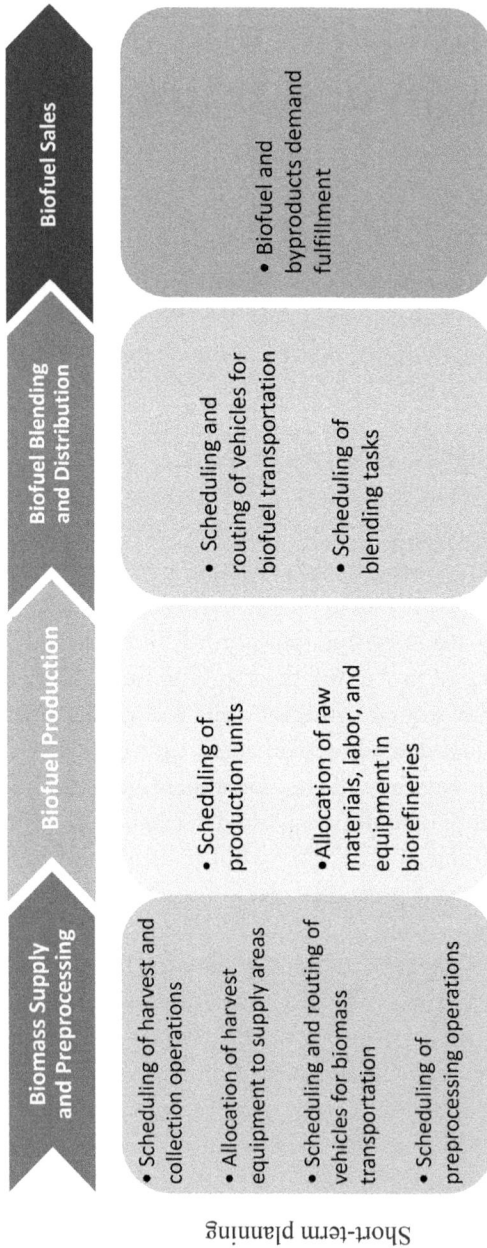

Figure 9.1 *An overview of the operational decisions in biomass supply chains.*

Biomass Supply and Preprocessing

- Scheduling of harvest and collection operations
- Allocation of harvest equipment to supply areas
- Scheduling and routing of vehicles for biomass transportation
- Scheduling of preprocessing operations

Biofuel Production

- Scheduling of production units
- Allocation of raw materials, labor, and equipment in biorefineries

Biofuel Blending and Distribution

- Scheduling and routing of vehicles for biofuel transportation
- Scheduling of blending tasks

Biofuel Sales

- Biofuel and byproducts demand fulfillment

Short-term planning

operations in biomass fields and allocate equipment and labor to them in an optimal manner (Malladi & Sowlati, 2018).

The pure scheduling problem is defined as assigning resources (e.g., labor, and machinery) to a set of operations over given time periods (Bochtis et al., 2013). In the biomass supply and logistics literature, the biomass harvesting and collection problem has often been cast as well-known combinatorial optimization problems such as traveling salesman problem (TSP), machine scheduling problem, and vehicle routing problem (VRP) with time windows aiming at determining the optimal schedules in several dispersed fields where a number of sequential harvesting tasks have to be performed using a set of machines. For example, Basnet, Foulds, and Wilson (2006) presented a harvesting scheduling method for harvesting rapeseed to produce canola oil based on a TSP, involving two sequential operations, namely swathing and threshing, and then developed a heuristic algorithm with the help of greedy heuristic and tabu search to obtain optimal solutions. Bochtis et al. (2013) and Orfanou et al. (2013) presented single-period machine scheduling models for the biomass harvesting and/or collection problem, in which a set of sequential operations (e.g., cutting, raking, baling, and loading) must be scheduled on different types of machines in each filed. The optimal schedule determines in which fields, in what sequence, and in which period of time each machine has to operate. Bochtis et al. (2013) formulated the scheduling problem of cotton residues collection operations involving two sequential tasks of baling and loading as a well-known flow shop problem with set up times, considering the objective function of minimizing the total completion time of the whole operation in all fields. Orfanou et al. (2013) extended the work of Bochtis et al. (2013) by taking into account multiple machinery systems and incorporating the corresponding cost of each schedule according to its time, which helps the decision maker by evaluating the relationship and tradeoff between time and cost for each potential schedule. Aguayo, Sarin, Cundiff, Comer, and Clark (2017) proposed static and dynamic scheduling problems for corn stover harvest based on an assumption that the harvest start time for each field can be known or unknown. They also assumed that one ethanol plant manages the harvesting and collection of corn stover from multiple fields and conveying stover bales to the plant over the harvest season. The proposed model formulated as a multiperiod mixed-integer programming model determines the minimum number of balers required in the harvest season, allocation of balers to the fields, as well as the routing of different types of balers per day, while minimizing the sum of the ownership cost of balers, corn stover collection cost, penalty cost of the late harvest, routing cost due to the movement of balers between fields, and makespan cost to obtain an optimal schedule with minimum length.

Operational scheduling of transportation operations in biomass supply chains focuses on the movement of biomass and biofuel between different locations throughout the chain over a short-term horizon ranging from one day to several weeks. The main short-term transportation decisions are determined by vehicle routing and scheduling problems that minimize the operating costs and/or the time that vehicles spent to transport biomass from supply areas to biorefineries as well as biofuel from biorefineries to demand zones. Recently, An (2019) presented a multiple-trips and visits VRP model for the daily transportation of corn stover bales stored as several satellite storage locations (SSLs) to a conversion facility using trucks and mobile loaders. In their model, it is assumed that mobile loaders travel between SSLs to support the loading operation of bales onto trucks. The developed mixed–integer linear programming (MILP) model determines the optimal daily schedule of trucks and loaders, including the required number of both types of machines, along with their routes, dispatching times, and synchronized working times at SSLs.

Biomass-to-biofuel supply chains are subjected to a wide range of uncertainties, which can be classified into short- and long-term uncertainties. The short-term uncertainties such as biomass price fluctuations refer to weekly or daily changes in influential parameters on the performance of biofuel supply chains, whereas the long-term uncertainties such as the evolution of biomass-to-biofuel conversion technology refer to changes that have a long-lasting influence over several years (Schütz, Tomasgard, & Ahmed, 2009). There are various sources of the short-term uncertainties, including biomass price and quality variations, disturbances and errors in operating conditions, biofuel demand fluctuation, variability in operating costs, and uncertainty in the completion time of supply chain activities (e.g., biomass harvest and collection, biomass and biofuel transportation, and biofuel productions) (Orfanou et al., 2013; Yue et al., 2014). Failure to hedge against such uncertainties in the biomass supply chain and logistics models may result in suboptimal or even infeasible operational decisions. Since operational decisions are adjusted frequently over a short period of time, they should be protected against short-term uncertainties. To do so, there is a need to employ an appropriate uncertainty handling approach that can capture short-term variations and dynamics encountered in biofuel supply chains, thereby ensuring the optimal flow of materials along the entire chain and continuous operations of the biorefineries and other supply chain components.

As stated in Chapter 6, stochastic programming and robust optimization have been the dominant mathematical programming approaches for decision-making under uncertainty, each having its own advantages and disadvantages. Stochastic programming typically optimizes the expected performance of the solution, but it entails the availability of the complete knowledge about the probability distributions of uncertain parameters, which is rarely available in practice. Besides, scenario-based stochastic programming approaches often employ a large set of scenarios to be responsive to different values the uncertain parameters may take, leading to computationally intractable problems (Pishvaee, Rabbani, & Torabi, 2011). On the other hand, in the robust optimization approaches, conventional set-based robust optimization approaches are classified as static robust optimization that assumes all decision variables are here-and-now, meaning that they must be determined before observing the actual realization of the uncertain parameters. This assumption, however, is very unrealistic in many dynamic real-worlds environments where some decisions can be made or revised when part of the uncertainty becomes known. In the biomass supply chain optimization problems, unlike the strategic decisions such as the location of biorefineries that needs to be decided at time zero and remains unchanged for several years, the operational decisions are typically made on a daily or weekly basis and can be revised based on the realizations of uncertain parameters in the previous period. To capture the sequential nature of decision-making in such problems, Ben-Tal, Goryashko, Guslitzer, and Nemirovski (2004) introduced adjustable robust optimization (ARO) that allows the recourse decisions to adjust themselves to the corresponding revealed information. This feature is particularly useful in immunizing operational level decisions of biomass supply chain optimization models, such as the scheduling and routing problems, because these decisions are made and occurred more frequently, and they can adjust themselves accordingly. Due to that fact that the uncertain parameters related to the operational decisions can often be described by a set of data samples, they can be handled by data-driven robust optimization approaches, which provides high-quality robust solutions when there are sufficient data samples that can be a good representative of the whole uncertainty domain. To deal with the uncertainty of input data, these approaches construct a flexible uncertainty set that accurately covers the region in which the past realizations of uncertainty data reside (Ning & You, 2019). The flexible uncertainty set provided by data-driven robust optimization encapsulates the most likely realizations of the uncertain parameters rather than every possible realization, thereby avoiding unnecessary conservatism associated with the classical uncertainty sets. In the following, harvest scheduling problem as one of the main operational-level problems in biofuel supply chains is presented, and uncertainty in biomass prices that introduce considerable risk into the decision-making process is dealt with using data-driven robust optimization.

9.2 Short-term corn stover harvest planning (a case study)

Corn stover including the husks, cobs, stalks, tassels, and leaves of the corn plant is one of the most promising feedstocks for bioethanol production due to its low cost, high carbohydrate content, and ready availability on a large scale (Yao et al., 2010). Optimal scheduling of stover harvest as one of the important aspects in planning the upstream of corn stover-to-ethanol supply chain is necessary for supplying sufficient quantities of stover in a reliable and cost-effective manner. Since it is not economical for corn growers to buy and maintain expensive corn stover harvest equipment that will be used only for a short harvest season, a third-party company is usually hired for stover harvesting, collection, and transportation to biorefineries using its own vehicles and machinery. Corn stover must be harvested and removed from the fields as quickly as possible after the grain harvest so that the farmers have enough time to prepare them for the next planting season (Ebadian, Sokhansanj, & Webb, 2017). The short harvesting window, along with the geographical dispersion of the fields, makes it challenging to determine the optimal harvesting plan that completes the harvesting process with minimum equipment requirements and total harvesting cost while simultaneously satisfying the harvest time window constraint for each field.

In the light of growing biomass demand for producing biofuels, harvest scheduling and planning models have received considerable attention in recent years. These models deal with the allocation of harvest equipment among corn fields to perform harvesting operations such as cutting, tedding, raking, baling, and hauling over a single period (Bochtis et al., 2013) or multiple periods (Aguayo et al., 2017), and their goal is to minimize the cost of harvesting (Karlsson, Rönnqvist, & Bergström, 2003) and/or the total time to complete harvesting (i.e., makespan) (Orfanou et al., 2013). This section develops a short-term stover harvest planning model that determines the number of balers required for stover harvest from several corn fields and assigns a sequence of fields to each baler, ensuring that stover harvest in each field is done within its allowable time window. The proposed model aims to maximize the total profit which is equal to revenues from selling stover to biorefineries minus the costs of harvesting. The selling price of stover that has a substantial, impact on the profit and harvesting decisions is difficult to estimate accurately. To cope with the uncertainty in the selling price, the harvest scheduling model is extended to a data–driven robust optimization version that is able to capture the complexity of the uncertainty data and to improve the robustness of the harvest schedule without imposing a significant cost.

9.2.1 Problem statement

After the harvesting operations of corn grain are done, the corn stover left on the field is ready to be harvested. The harvest of stover must be finished before winter when the ground may be wet, muddy, or covered with snow (Ebadian et al., 2017). Therefore, a harvest time window, which is delimited by ready and due dates, is considered for each field to ensure the removal of stover within an appropriate time period. The scheduling problem considered in this study seeks to determine the optimal number of balers required for stover harvest in several geographically distributed fields along with the optimal schedule for each baler, ensuring that the harvest of each field is completed within its corresponding time window and that the net profit from harvesting and selling stover is maximized. The optimal schedule specifies which fields should be served by each baler and in what order each baler should visit its assigned fields.

The following assumptions are also made to establish the harvest scheduling model:

- There a machinery depot where balers depart from when they are assigned a field and return to after the harvest of stover in the assigned fields is completed.
- At the end of each time stage, balers are allowed to stay at their current field, move to another field, or return to the depot.
- As the harvest of the whole available stover in some fields may need an additional baler operating at less than full capacity, a portion of stover can be left unharvested in each field.
- Corn stover is harvested as square bales that offer the advantages of ease of transport, greater bulk density, and higher baler productivity compared to round bales (Grisso, Mccullough, Cundiff, & Judd, 2013).
- Since the objective of this study is to determine short-term operational scheduling decisions for corn stover harvesting, each time period in the

model is defined as one day to be able to handle variations in daily harvesting operations.

The selling price of stover is one of the key parameters influencing the net profit of corn growers or third-party companies responsible for harvesting stover from corn fields and selling it to biorefineries. The price of stover is dependent on the quality of stover (i.e., ash and moisture content) which itself is a function of climate conditions, soil properties, harvesting time, and many other factors (Thompson & Tyner, 2014). Therefore, it is difficult to estimate the value of stover price accurately. The previous stover harvest planning models disregard uncertainty in their input parameters, resulting in suboptimal or even infeasible schedules in practice.

Motivated by the above practical concern, this study hedges against uncertainty in the selling price of stover using the data-driven robust optimization approach developed by (Shang, Huang, & You, 2017). The conventional robust optimization approaches are based on uncertainty sets with fixed geometric shapes that are unable to adjust themselves to the structure of uncertainty data (Ning & You, 2019). Therefore, they cannot capture possible correlation and asymmetry in the prices of stover harvested from different fields, particularly those fields that are located in the vicinity of each other and produce stover of similar quality due to the same weather conditions. Data-driven robust optimization approaches, on the other hand, construct uncertainty sets from uncertainty data samples while accurately covering the region in which possible uncertainty realizations may reside and implicitly taking into account the correlation and asymmetry of uncertainty data (Mohseni & Pishvaee, 2019). This flexibility enables us to find the smallest possible uncertainty set that encompasses all uncertainty realizations with no superfluous coverage, resulting in a significantly lower cost of robustness compared to the conventional robust optimization approaches that employ uncertainty sets whose structure is fixed a priori.

To illustrate the optimal schedule determined by the model, we consider a harvest schedule problem with five fields where stover should be harvested within given time windows. As shown in Fig. 9.2, two balers can complete the stover harvest in the fields in 6 days. The optimal schedule for baler 1 is $\left[3^1 - 4^2 - 4^3 - 4^4 - 1^5 - 1^6\right]$, where f^t represents that the baler works in field f during period t. Baler 2 is responsible for harvesting fields 5, 2 and 1 with a schedule as $\left[5^1 - 5^2 - 2^3 - 2^4 - 1^5\right]$. It should be noted the self-loops in Fig. 9.2 mean that the operation of a baler in a field continues for successive time periods.

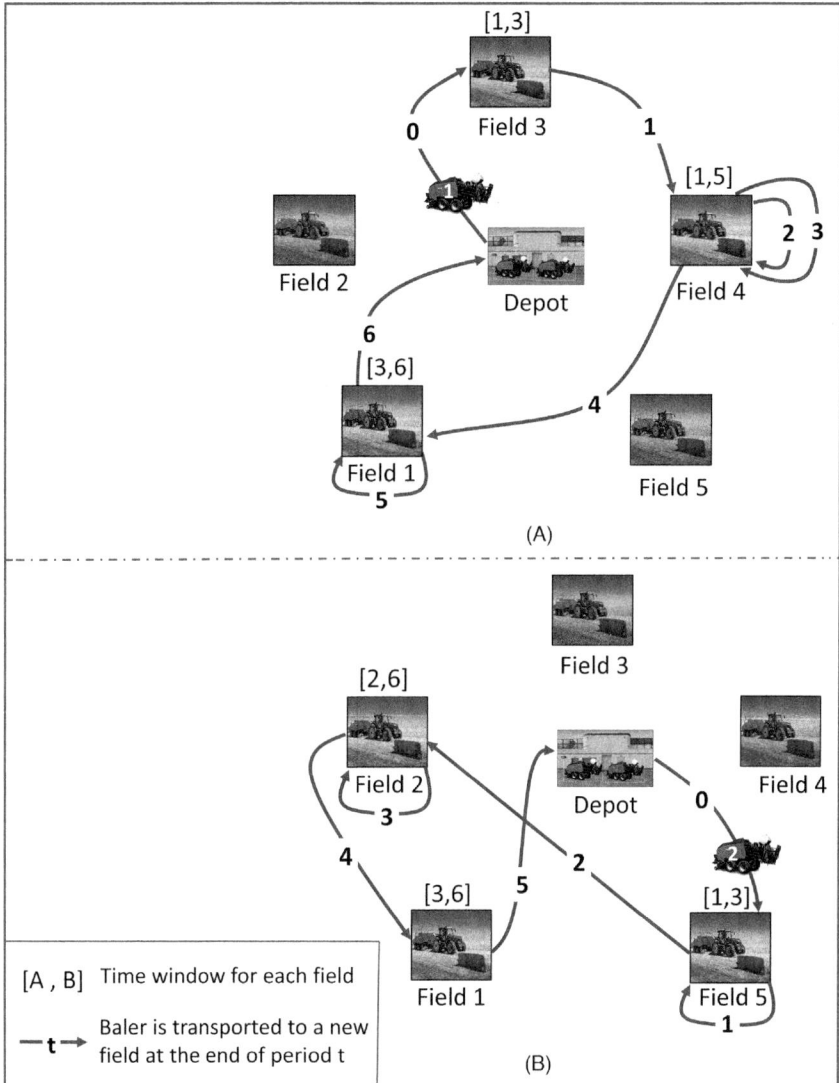

Figure 9.2 *The optimal schedules for balers required for stover harvest from five fields.* (A) The schedule for baler 1 and (B) the schedule for baler 2.

9.2.2 Mathematical model

The harvest scheduling problem described in the previous section is formulated as a MILP model. A list of indices, parameters, and decision variables required to formulate the model is given in Table 9.1.

Table 9.1 Indices, parameters, and decision variables used in the scheduling model.

Sets

F	Set of fields, indexed by f
F_0	Set of fields including the depot, indexed by f
T	Set of time periods, indexed by t

Parameters

q	Harvesting capacity of a baler
A_f	Total amount of stover available in field f
Max_f	Maximum possible harvest level in field f in each period
Min_f	Minimum possible harvest level in field f in each period
R_f	Ready date of the stover harvest in field f
D_f	Due date of the stover harvest in field f
p_f	Selling price of stover harvested from field f
FC	Fixed cost of a baler
VC	Variable cost of harvesting one unit of stover
$TC_{i,j}$	Cost of transporting a baler from location i to location j

Continuous variables $(0\ to+\infty)$

x_f^t	Amount of stover harvested from field f in period t
S_f	Starting period of the stover harvest in field f
E_f	Ending period of the stover harvest in field f

Integer variables $(0\ to+\infty)$

n_f^t	Number of balers allocated to field f in period t
$Z_{f,i}^t$	Number of balers transported from location i to location j at the end of period t
N	Number of balers required during the harvest season
Y_f^t	Binary variable, equal to 1 if field f is harvested in period t; 0 otherwise

9.2.2.1 Objective function

The objective function (9.1) is to maximize the total profit throughout the entire planning horizon, which is equal to revenues from selling corn stover to biorefineries minus the harvesting cost consisting of three main components. The first is the variable cost of harvesting including the costs of labor, fuel, and materials (such as net wrap for binding the bales). The second component is the fixed cost of balers used during the harvest season and the last one is the cost of transporting balers between various locations:

$$\text{Max} \sum_{f\in F}\sum_{t\in T}\tilde{p}_f.x_f^t - \sum_{f\in F}\sum_{t\in T}VC.x_f^t - FC.N - \sum_{i\in F_0}\sum_{j\in F_0}\sum_{t\in T}TC_{i,j}.Z_{i,j}^t \quad (9.1)$$

9.2.2.2 Constraints

The number of balers needed during period t to harvest the stover in field f is given by constraint (9.2). This nonlinear constraint can be recast in a linear form as in (9.3). The total number of balers required during the harvest season is determined by constraint (9.4). Constraint (9.5) ensures that, in each field, the sum of stover harvested in different periods is not greater than the total available stover. Constraint (9.6) ensures that if field f is harvested in period t, the amount of stover harvested is between the minimum and maximum possible harvest levels. Constraint (9.7) ensures that the harvest of stover in field f cannot start before its ready date. Constraint (9.8) ensures that field f can be harvested if at least one baler is allocated to the field. For each field, the starting and ending periods of the harvest are determined by constraints (9.9) and (9.10), respectively. Constraint (9.11) ensures the completion of the harvest in each field by its due date. Constraint (9.12) is a logical constraint ensuring that the ending time of the harvest in each field is not earlier than its starting time. Constraint (9.13) ensures that field f is continuously harvested from its starting to its ending time period. Constraint (9.14) ensures that the number of balers remaining in field f at the end of period $t-1$ (i.e., $Z_{i,f}^{t-1}$) plus the number of new blares transported from other fields to field f at the end of period $t-1$ (i.e., $\sum_i Z_{i,f}^{t-1}, i \neq f$) must be equal to the number of balers working in field f during period t. Constraint (9.15) ensures that the number of balers working in field f during period t must be equal to the number of balers leaving the field and moving to a new field at the end of period t plus the number of balers staying in field f for the next period. The fact that all balers must depart from and return to the depot is expressed by constraints (9.16) and (9.17).

$$n_f^t = \left\lceil \frac{x_f^t}{q} \right\rceil \quad \forall f \in F, t \in T \tag{9.2}$$

$$\frac{x_f^t}{q} \leq n_f^t \leq \frac{x_f^t}{q} + 0.99 \quad \forall f \in F, t \in T \tag{9.3}$$

$$N \geq \sum_{f \in F} n_f^t \quad \forall t \in T \tag{9.4}$$

$$\sum_{t \in T} x_f^t \leq A_f \quad \forall f \in F \tag{9.5}$$

$$Min_f . Y_f^t \leq x_f^t \leq Max_f . Y_f^t \quad \forall f \in F, t \in T \tag{9.6}$$

$$Y_f^t = 0 \quad \forall f \in F, t \le R_f - 1 \tag{9.7}$$

$$Y_f^t \le n_f^t \quad \forall f \in F, t \in T \tag{9.8}$$

$$S_f \le t + |T|\left(1 - Y_f^t\right) \quad \forall f \in F, t \in T \tag{9.9}$$

$$E_f \ge t.Y_f^t \quad \forall f \in F, t \in T \tag{9.10}$$

$$E_f \le D_f \quad \forall f \in F \tag{9.11}$$

$$S_f \le E_f \quad \forall f \in F \tag{9.12}$$

$$\sum_{t \in T} Y_f^t = E_f - S_f + 1 \quad \forall f \in F \tag{9.13}$$

$$\sum_{i \in F_0} Z_{i,f}^{t-1} = n_f^t \quad \forall f \in F, t \in T \tag{9.14}$$

$$n_f^t = \sum_{i \in F_0} Z_{f,i}^t \quad \forall f \in F, t \in T \tag{9.15}$$

$$\sum_{f \in F} \sum_{t \in T} Z_{0,f}^t = N \tag{9.16}$$

$$\sum_{f \in F} \sum_{t \in T} Z_{f,0}^t = N \tag{9.17}$$

9.2.2.3 Data-driven robust optimization model

As explained in detail in Chapter 6, the data–driven robust optimization approach proposed by Shang et al. (2017) is comprised of several main steps, which must be taken to develop the robust counterpart of the proposed harvest scheduling model.

Step 1: A set of data samples $\left\{p^i\right\}_{i=1}^N$ is collected as possible realizations of corn stover prices. Each sample is described by a F-dimensional vector, $p = \left[p_1, p_2, \ldots, p_F\right]^T$, which corresponds to the selling price of stover harvested in different fields.

Step 2: The estimated covariance matrix Σ and the weighting matrix Q are calculated by the following equation:

$$Q = \Sigma^{-\frac{1}{2}} = \left[\frac{1}{N-1}\left(\sum_{i=1}^N p^i(p^i)^T - \left(\sum_{i=1}^N p^i\right)\left(\sum_{i=1}^N p^i\right)^T\right)\right]^{-\frac{1}{2}} \tag{9.18}$$

Step 3: Using the kernel (9.19) tuned according to relation (9.20), the quadratic programming problem (9.21) is solved:

$$K\left(p^i, p^j\right) = \sum_{k=1}^{F} l_k - \left\|Q\left(p^i - p^j\right)\right\|_1 \tag{9.19}$$

$$l_k > \max_{i=1}^{N} q_k^T p^i - \min_{i=1}^{N} q_k^T p^i \tag{9.20}$$

$$\max \sum_{i=1}^{N} \alpha_i K\left(p^i, p^i\right) - \sum_{i=1}^{N}\sum_{j=1}^{N} \alpha_i \alpha_j K\left(p^i, p^j\right)$$
$$\text{s.t.} \quad \sum_{i=1}^{N} \alpha_i = 1 \tag{9.21}$$
$$0 \le \alpha_i \le V$$

Problem (9.21) is the Wolfe dual form of the support vector (SV) clustering model that is aimed at discovering the smallest sphere encapsulating data samples (for details the reader is referred to Chapter 6). The adjustable parameter V controls the number of data samples allowed to be located outside the sphere, thereby regulating the size of the sphere (i.e., the uncertainty set) and the conservatism level of the robust solution.

Step 4: Based on the optimal values of α_i, data samples with $0 < \alpha_i < V$ that are named SVs reside on the boundary of the sphere, data samples with $\alpha_i = V$ that are named bounded SVs (BSV) lie outside the sphere, and the rest samples that are named inlier data (ID) samples are enclosed within the sphere.

Step 5: The uncertainty set that encompasses the region where the future values of stover prices are more likely to be located is constructed as follows:

$$U\left(p\right) = \left\{ p \left| \sum_{i \notin ID} \alpha_i \left\|Q\left(p - p^i\right)\right\|_1 \le \sum_{i \notin ID} \alpha_i \left\|Q\left(p^{i'} - p^i\right)\right\|_1 \quad i' \in SV \right. \right\} \tag{9.22}$$

which is equivalently rewritten as,

$$U\left(p\right) = \left\{ p \left| \begin{array}{l} \exists z_i, i \notin ID, \quad s.t. \\ \sum_{i \notin ID} \alpha_i \cdot z_i^T 1 \le \Omega \\ -z_i \le Q\left(p - p^i\right) \le z_i, i \notin ID \end{array} \right. \right\} \tag{9.23}$$

where $\Omega = \sum_{i \notin ID} \alpha_i \left\|Q\left(p^{i'} - p^i\right)\right\|_1, i' \in SV$, and z_i is a F-dimensional auxiliary variable.

Step 6: The uncertain parameters in the objective function are transformed into the constraints by introducing the auxiliary variable R, resulting in the following reformulation (in vector form):

$$\text{Max} \quad R + vc^T x - fc^T n - tc^T z$$
$$\text{s.t.} \quad R \leq \tilde{p}^T x \tag{9.24}$$
$$\text{Constraints} \ (9.3)-(9.17)$$

Step 7: Under the data-driven uncertainty set (DDUS) (9.23), the robust counterpart of the first constraint of model (9.24) is formulated as follows:

$$R \leq \min_{\tilde{p} \in U} \tilde{p}^T x \tag{9.25}$$

Step 8: Replacing the inner minimization problem with its dual equivalent, the following constraint set is obtained.

$$\begin{cases} R \leq \displaystyle\sum_{i \notin ID} (n_i - m_i)^T Qu^i - \lambda\Omega \\[2mm] \displaystyle\sum_{i \notin ID} Q(n_i - m_i) - x = 0 \\[2mm] m_i + n_i = \lambda \cdot \alpha_i \cdot 1 \quad \forall i \notin ID \\[2mm] n_i, m_i \in R_+^n \quad x, \lambda \geq 0 \end{cases} \tag{9.26}$$

where m_i, n_i, and λ are the dual variables.

9.2.3 Results and discussion

In this section, the performance of the proposed model is evaluated through a harvest scheduling problem aiming to complete the harvest of 15 corn fields within the planning horizon of 20 days. Except for the data related to stover selling prices, all other data required to run the model can be found in Aguayo et al. (2017). For modeling the realizations of stover price fluctuations, the average price in each field is first estimated based on the four quality grades (four price levels) defined by Thompson and Tyner (2014), and a set of 300 random samples is generated around the average prices using a correlated Gaussian distribution to incorporate possible correlations and asymmetries embodied in the price data in the different fields.

To testify the superiority of the DDUS, the harvest scheduling problem is also solved using the integrated box-polyhedral uncertainty set (BPUS) as the most commonly adopted conventional uncertainty set. For solving the robust model under this integrated uncertainty set, it is assumed that the stover price varies between the minimum and maximum value of the generated samples, and the uncertainty budget that controls the degree of conservatism of the robust solution takes values between 0 and 15. The robust counterpart of the scheduling model under the BPUS is given in (9.27). The reader is referred to Chapter 6 for more information on how to formulate the robust counterpart model under this uncertainty set. All computational experiments are carried out on a desktop PC equipped with an Intel Core i7-950 processor operating at 3.40 GHz. The quadratic problem (9.21) is solved using CVX, a convex optimization package implemented in MATLAB (Grant & Boyd, 2014), and the deterministic and robust harvest scheduling problems are solved using CPLEX 12.5 solver in GAMS software.

$$
\begin{aligned}
\text{Max} \quad & R + \sum_{f \in F} \sum_{t \in T} VC . x_f^t - FC . N - \sum_{i \in F_0} \sum_{j \in F_0} \sum_{t \in T} TC_{i,j} . Z_{i,j}^t \\
\text{s.t.} \quad & R \le \sum_{f \in F} \sum_{t \in T} p_f . x_f^t - \Gamma z - \sum_{f \in F} \sum_{t \in T} \omega_f^t \qquad (9.27) \\
& z + \omega_f^t \ge \hat{p}_f . x_f^t \quad \forall f, t \\
& \text{Constraints } (9.3) - (9.17)
\end{aligned}
$$

where Γ is the uncertainty budget, and $\hat{p}_f$ is the maximum deviation of the stover price in each field from its nominal value (p_f).

9.2.3.1 Comparison of harvest profits obtained by different models

As stated earlier, the conservatism of the robust models with the DDUS and BPUS are adjusted by the parameter V in (9.21) and the uncertainty budget Γ, respectively. To compare the performance of the robust harvest scheduling models with the DDUS and BPUS, they are first solved under different values of their corresponding adjustable parameters. Then, the robustness of the obtained solutions is evaluated using the Monte–Carlo simulation. For this simulation, 500 data samples are generated and the robustness levels provided by the DDUS and BPUS are measured based on the number of samples that are covered by them for different values of V and Γ. To make a fair comparison, it is assumed that 90% of the samples follow the correlated Gaussian distribution that was used to describe the stover prices, and the other 10% are randomly distributed within the variation range of the stover prices to stimulate the anomalous behavior of the stover prices.

According to the results of Fig. 9.3, the deterministic model in which all parameters are set at their nominal values results in a profit of 238,290.

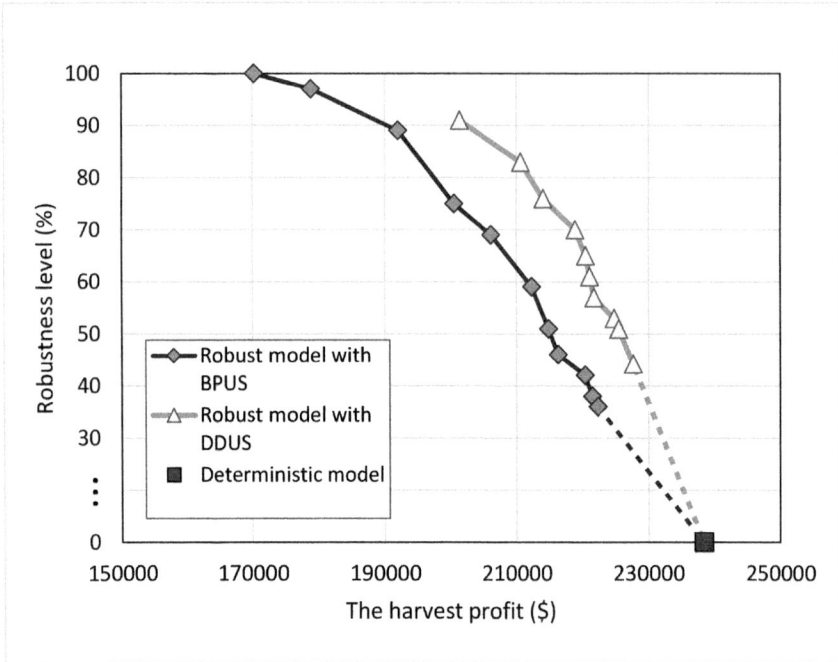

Figure 9.3 *The harvest profits generated by the deterministic model and the robust models with the BPUS and DDUS.*

Although the harvest schedule determined by the deterministic model generates a higher profit compared with those determined by the robust models, it is not sufficiently reliable to be implemented in practice because it has no protection against uncertainty and becomes suboptimal if the prices of stover slightly deviate from their nominal value. The robust models, on the other hand, overcome this problem by improving the robustness of the schedule in the face of uncertainty, but this comes at the expense of a decrease in the expected profit. The difference between the profit of the deterministic and robust models, which can be interpreted as the cost of robustness, is lower for the DDUS. This means that the robust model with DDUS sacrifices less optimality for the robustness compared to the robust model with the BPUS. To ensure the robustness level of 90%, for example, the robustness cost incurred by the robust model with the DDUS is 36,990, which is about 20% lower than that incurred by the robust model with the BPUS. Therefore, the proposed robust model reduces the cost of robustness by employing the uncertainty set constructed based on the geometrical structure of the underlying uncertainty data.

The results also show that the robust model with the BPUS produces robust solutions with the highest robustness levels, which are guaranteed to be robust against almost all the possible realizations of the uncertain parameters within their variation range. These solutions, however, are overly conservative in practice and impose a huge cost on the company responsible for harvesting corn stover. While the BPUS encapsulates every possible realization, even those with a very low probability, the DDUS adjusts itself such that only the likely values of the uncertain parameters are covered, thereby making a reasonable trade-off between robustness and its cost.

9.2.3.2 Comparison of harvest schedules obtained by different models

In addition to the deterioration of the objective function value (i.e., the harvest profit), it is of great importance to see how the solution robustness is reflected in the harvest schedules determined by the robust model. Fig. 9.4 provides a comparison between the optimal schedules generated using the deterministic model, the robust model with the BPUS, and the robust model with the DDUS. For a fair comparison, the regulation parameters in both the uncertainty sets are adjusted such that the induced robustness levels are almost the same (around 90%). It should be noted that due to the discrete nature of regulation of the coverage of uncertainty data in the robust models, it is difficult to have exactly identical robustness levels.

As can be seen from Fig. 9.4, the deterministic and robust models give the same results regarding the number of balers required for harvesting and the sequence of fields to be harvested by each baler. This indicates that these decisions are robust in the face of uncertainty in the price of stover and are not invalidated by the fluctuations that might occur in the price of stover. However, the number of days devoted to the stover harvest in each field is determined differently. The deterministic model leads to more harvesting days and less idle periods compared to the robust models. In view of the fact that the deterministic model only accounts for the nominal values of the stover prices, the revenue from selling stover harvested in every field is often higher than the cost of harvesting; therefore, almost all the available stover in the different fields is harvested to maximize the total profit. On the contrary, in the robust models that hedge against the worst-case values of stover prices, the harvesting in some fields is less profitable or even unprofitable, and a reasonable strategy is to leave a fraction of the available stover unharvested.

The results also show that the robust model with BPUS reduces the harvest of stover in the fields with lower harvest levels (lower stover price). The

(A) Deterministic model

	Baler 1	Baler 2	Baler 3
Day 1	Field 3	Field 14	Field 6
Day 2	Field 3	Field 14	Field 6
Day 3	Field 5	Field 2	Field 2
Day 4	Field 5	Field 2	Field 2
Day 5		Field 2	
Day 6	Field 12	Field 12	Field 4
Day 7	Field 12	Field 12	Field 4
Day 8	Field 9	Field 15	Field 4
Day 9	Field 9	Field 15	
Day 10		Field 8	Field 7
Day 11	Field 1	Field 8	Field 7
Day 12	Field 1	Field 1	Field 1
Day 13	Field 1	Field 1	Field 1
Day 14			Field 10
Day 15			Field 10
Day 16	Field 13	Field 13	Field 13
Day 17	Field 13	Field 13	Field 13
Day 18	Field 13	Field 11	Field 11
Day 19		Field 11	Field 11
Day 20			

(B) Robust model with BPUS

	Baler 1	Baler 2	Baler 3
Day 1	Field 3	Field 14	Field 6
Day 2	Field 3	Field 2	Field 6
Day 3	Field 5	Field 2	Field 2
Day 4		Field 2	Field 2
Day 5			
Day 6	Field 12	Field 12	Field 4
Day 7	Field 12	Field 12	
Day 8	Field 9	Field 15	
Day 9		Field 8	
Day 10		Field 8	Field 7
Day 11	Field 1	Field 1	Field 1
Day 12	Field 1	Field 1	Field 1
Day 13	Field 1		
Day 14			Field 10
Day 15			
Day 16	Field 13	Field 13	Field 13
Day 17	Field 13	Field 13	Field 13
Day 18	Field 13	Field 11	Field 11
Day 19		Field 11	Field 11
Day 20			

(C) Robust model with DDUS

	Baler 1	Baler 2	Baler 3
Day 1	Field 3	Field 14	Field 6
Day 2	Field 3	Field 14	Field 6
Day 3	Field 5	Field 2	Field 2
Day 4		Field 2	Field 2
Day 5			
Day 6	Field 12	Field 12	Field 4
Day 7		Field 12	Field 4
Day 8	Field 9	Field 15	
Day 9	Field 9	Field 15	
Day 10		Field 8	Field 7
Day 11		Field 8	Field 7
Day 12	Field 1	Field 1	Field 1
Day 13	Field 1	Field 1	Field 1
Day 14			Field 10
Day 15			Field 10
Day 16	Field 13	Field 13	Field 13
Day 17	Field 13	Field 13	Field 11
Day 18		Field 11	Field 11
Day 19			
Day 20			

Figure 9.4 *The optimal harvest schedules generated by the deterministic model and the robust models with the BPUS and DDUS.* (A) Deterministic model, (B) robust model with BPUS, and (C) robust model with DDUS.

possible reason is that the BPUS assumes there is no correlation between the prices of stover in the different fields and tends to reduce the harvesting in the fields with the lowest stover prices under the worst-case condition. However, since the DDUS considers correlations among the prices in the different fields, all the fields might be subjected to the reduction of harvesting when the worst-case scenarios for stover prices are materialized.

9.3 Conclusions

Operational decisions in the supply chain are typically made weekly, daily, or even hourly, based on the type and characteristics of the chain. Unlike the strategic decisions, operational decisions can be altered and revised more frequently with regard to different internal and external factors influencing the performance of the supply chain. This chapter presents the operational decisions that are made at different stages of biomass-to-biofuel supply chains and discusses the most recognized types of operational uncertainties threatening the biomass procurement stage.

In the light of growing biomass demand for producing biofuels, biomass harvest scheduling and planning models have received considerable attention in recent years. These models deal with the allocation of harvest equipment to corn fields to perform harvesting operations. In order to address harvest scheduling problem, a short-term corn stover harvest planning model is proposed in this chapter that determines the number of balers required for stover harvest from several corn fields and assigns a sequence of fields to each baler, ensuring that stover harvest in each field is done within its allowable time window. The proposed model aims to maximize the total profit which is equal to revenues from selling stover to biorefineries minus the costs of harvesting. To hedge against uncertainty in the selling price of stover, the harvest scheduling model is extended to a data-driven robust optimization version that is able to capture the complexity of the uncertainty data and to improve the robustness of the harvest schedule without imposing a significant cost. Finally, the performance of the proposed model is evaluated through a harvest scheduling problem aiming to complete the harvest of 15 corn fields within the planning horizon of 20 days. The results show that the proposed data-driven scheduling model returns more profit that the conventional robust optimization models while ensuring the same level of protection against uncertainty.

References

Aguayo, M. M., Sarin, S. C., Cundiff, J. S., Comer, K., & Clark, T. (2017). A corn-stover harvest scheduling problem arising in cellulosic ethanol production. *Biomass and Bioenergy*, *107*, 102–112.

An, H. (2019). Optimal daily scheduling of mobile machines to transport cellulosic biomass from satellite storage locations to a bioenergy plant. *Applied Energy*, *236*, 231–243.

Balaman, Ş. Y. (2019). Chapter 6 - Basics of decision-making in design and management of biomass-based production chains. In Ş. Y. Balaman (Ed.), *Decision-making for biomass-based production chains*Academic Press.

Basnet, C. B., Foulds, L. R., & Wilson, J. M. (2006). Scheduling contractors' farm-to-farm crop harvesting operations. *International Transactions in Operational Research*, *13*, 1–15.

Ben-Tal, A., Goryashko, A., Guslitzer, E., & Nemirovski, A. (2004). Adjustable robust solutions of uncertain linear programs. *Mathematical Programming*, *99*, 351–376.

Bochtis, D., Dogoulis, P., Busato, P., Sørensen, C., Berruto, R., & Gemtos, T. (2013). A flow-shop problem formulation of biomass handling operations scheduling. *Computers and Electronics in Agriculture*, *91*, 49–56.

Ebadian, M., Sokhansanj, S., & Webb, E. (2017). Estimating the required logistical resources to support the development of a sustainable corn stover bioeconomy in the USA. *Biofuels, Bioproducts and Biorefining*, *11*, 129–149.

Grant, M., & Boyd, S. (2014). *CVX: Matlab software for disciplined convex programming, version 2.1*.

Grisso, R. D., Mccullough, D., Cundiff, J. S., & Judd, J. D. (2013). Harvest schedule to fill storage for year-round delivery of grasses to biorefinery. *Biomass and Bioenergy*, *55*, 331–338.

Karlsson, J., Rönnqvist, M., & Bergström, J. (2003). Short-term harvest planning including scheduling of harvest crews. *International Transactions in Operational Research*, *10*, 413–431.

Malladi, K. T., & Sowlati, T. (2018). Biomass logistics: A review of important features, optimization modeling and the new trends. *Renewable and Sustainable Energy Reviews*, *94*, 587–599.

Mohseni, S., & Pishvaee, M. S. (2019). Data-driven robust optimization for wastewater sludge-to-biodiesel supply chain design. *Computers & Industrial Engineering*, *139*, 105944.

Ning, C., & You, F. (2019). Optimization under uncertainty in the era of big data and deep learning: When machine learning meets mathematical programming. *Computers & Chemical Engineering*, *125*, 434–448.

Orfanou, A., Busato, P., Bochtis, D., Edwards, G., Pavlou, D., Sørensen, C. G., & Berruto, R. (2013). Scheduling for machinery fleets in biomass multiple-field operations. *Computers and Electronics in Agriculture*, *94*, 12–19.

Pishvaee, M. S., Rabbani, M., & Torabi, S. A. (2011). A robust optimization approach to closed-loop supply chain network design under uncertainty. *Applied Mathematical Modelling*, *35*, 637–649.

Schütz, P., Tomasgard, A., & Ahmed, S. (2009). Supply chain design under uncertainty using sample average approximation and dual decomposition. *European Journal of Operational Research*, *199*, 409–419.

Shang, C., Huang, X., & You, F. (2017). Data-driven robust optimization based on kernel learning. *Computers & Chemical Engineering*, *106*, 464–479.

Thompson, J. L., & Tyner, W. E. (2014). Corn stover for bioenergy production: Cost estimates and farmer supply response. *Biomass and Bioenergy*, *62*, 166–173.

Yao, L., Yue, J., Zhao, J., Dong, J., Li, X., & Qu, Y. (2010). Application of acidic wastewater from monosodium glutamate process in pretreatment and cellulase production for bioconversion of corn stover–Feasibility evaluation. *Bioresource Technology*, *101*, 8755–8761.

Yue, D., You, F., & Snyder, S. W. (2014). Biomass-to-bioenergy and biofuel supply chain optimization: Overview, key issues and challenges. *Computers & Chemical Engineering*, *66*, 36–56.

Index

Note: Page numbers followed by "f" indicate figures, "t" indicate tables.